THE BOOK OF THE
HONEY BEE

BY

CHARLES HARRISON

WITH NUMEROUS ILLUSTRATIONS
CHIEFLY AFTER PHOTOGRAPHS BY
JOHN C. DOUGLAS

British Library Cataloguing-in-Publication Data
A catalogue record for this book is available from the
British Library

Bee Keeping

Beekeeping (or apiculture, from Latin: *apis* 'bee') is quite simply, the maintenance of honey bee colonies. A beekeeper (or apiarist) keeps bees in order to collect their honey and other products that the hive produces (including beeswax, propolis, pollen, and royal jelly), to pollinate crops, or to produce bees for sale to other beekeepers. A location where bees are kept is called an apiary or 'bee yard.' Depictions of humans collecting honey from wild bees date to 15,000 years ago, and efforts to domesticate them are shown in Egyptian art around 4,500 years ago. Simple hives and smoke were used and honey was stored in jars, some of which were found in the tombs of pharaohs such as Tutankhamun.

The beginnings of 'bee domestication' are uncertain, however early evidence points to the use of hives made of hollow logs, wooden boxes, pottery vessels and woven straw baskets. On the walls of the sun temple of Nyuserre Ini (an ancient Egyptian Pharo) from the Fifth Dynasty, 2422 BCE, workers are depicted blowing smoke into hives as they are removing honeycombs. Inscriptions detailing the production of honey have also been found on the tomb of Pabasa (an Egyptian nobleman) from the Twenty-sixth Dynasty (c. 650 BCE), depicting pouring honey in jars and cylindrical hives. Amazingly though, archaeological finds relating to beekeeping have been discovered at Rehov, a Bronze and Iron Age archaeological site in the Jordan Valley, Israel.

Thirty intact hives, made of straw and unbaked clay, were discovered in the ruins of the city, dating from about 900 BCE. The hives were found in orderly rows, three high, in a manner that could have accommodated around 100 hives, held more than 1 million bees and had a potential annual yield of 500 kilograms of honey and 70 kilograms of beeswax!

It wasn't until the eighteenth century that European understanding of the colonies and biology of bees allowed the construction of the moveable comb hive so that honey could be harvested without destroying the entire colony. In this 'Enlightenment' period, natural philosophers undertook the scientific study of bee colonies and began to understand the complex and hidden world of bee biology. Preeminent among these scientific pioneers were Swammerdam, René Antoine Ferchault de Réaumur, Charles Bonnet and the Swiss scientist Francois Huber. Huber was the most prolific however, regarded as 'the father of modern bee science', and was the first man to prove by observation and experiment that queens are physically inseminated by drones outside the confines of hives, usually a great distance away. Huber built improved glass-walled observation hives and sectional hives that could be opened like the leaves of a book. This allowed inspecting individual wax combs and greatly improved direct observation of hive activity. Although he went blind before he was twenty, Huber employed a secretary, Francois Burnens, to make daily observations, conduct careful experiments, and keep accurate notes for more than twenty years.

Early forms of honey collecting entailed the destruction of the entire colony when the honey was harvested. The wild hive was crudely broken into, using smoke to suppress the bees, the honeycombs were torn out and smashed up — along with the eggs, larvae and honey they contained. The liquid honey from the destroyed brood nest was strained through a sieve or basket. This was destructive and unhygienic, but for hunter-gatherer societies this did not matter, since the honey was generally consumed immediately and there were always more wild colonies to exploit. It took until the nineteenth century to revolutionise this aspect of beekeeping practice – when the American, Lorenzo Lorraine Langstroth made practical use of Huber's earlier discovery that there was a specific spatial measurement between the wax combs, later called *the bee space*, which bees do not block with wax, but keep as a free passage. Having determined this bee space (between 5 and 8 mm, or 1/4 to 3/8"), Langstroth then designed a series of wooden frames within a rectangular hive box, carefully maintaining the correct space between successive frames, and found that the bees would build parallel honeycombs in the box without bonding them to each other or to the hive walls.

Modern day beekeeping has remained relatively unchanged. In terms of keeping practice, the first line of protection and care – is always sound knowledge. Beekeepers are usually well versed in the relevant information; biology, behaviour, nutrition - and also wear protective clothing. Novice beekeepers commonly wear gloves and a hooded suit or hat and veil, but some experienced beekeepers elect not to use gloves because they inhibit delicate manipulations. The face and neck are the most important areas to protect (as a sting here will lead to much more pain and swelling than a sting elsewhere), so most beekeepers wear at least a veil. As an interesting note, protective clothing is generally white, and of a smooth material. This is because it provides the maximum differentiation from the colony's natural predators (bears, skunks, etc.), which tend to be dark-coloured and furry. Most beekeepers also use a 'smoker'—a device designed to generate smoke from the incomplete combustion of various fuels. Smoke calms bees; it initiates a feeding response in anticipation of possible hive abandonment due to fire. Smoke also masks alarm pheromones released by guard bees or when bees are squashed in an inspection. The ensuing confusion creates an opportunity for the beekeeper to open the hive and work without triggering a defensive reaction.

Such practices are generally associated with rural locations, and traditional farming endeavours. However, more recently, urban beekeeping has emerged; an attempt to revert to a less industrialized way of obtaining honey by utilizing small-scale colonies that pollinate urban gardens. Urban apiculture has undergone a renaissance in the first decade of the twenty-first century, and urban beekeeping is seen by many as a growing trend; it has recently been legalized in cities where it was previously banned. Paris, Berlin, London, Tokyo, Melbourne and Washington DC are among beekeeping cities. Some have found that 'city bees' are actually healthier than 'rural bees' because there are fewer pesticides and greater biodiversity. Urban bees may fail to find forage, however, and homeowners can use their landscapes to help feed local bee populations by planting flowers that provide nectar and pollen. As is evident from this short introduction, 'Bee-Keeping' is an incredibly ancient practice. We hope the current reader is inspired by this book to be more 'bee aware', whether that's via planting appropriate flowers, keeping bees or merely appreciating! Enjoy.

CONTENTS

CONTENTS

LIST OF ILLUSTRATIONS

EDITOR'S NOTE

FROM the time of Virgil to our own day bee-keeping has been the branch of husbandry which has peculiarly appealed to the temperament of meditative man. Nor does the charm of the "bee-loud glade" grow less with our increasing knowledge of the life that runs its marvellous course within the walls of the hive. What patient labour has been spent in the gathering of this knowledge is known but to few. Swammerdam, Réaumur and, greatest and most patient of them all, the blind Huber—these are but three out of hundreds of names that are worthy of honour in the history of the science.

Of all recent events that have drawn towards the hive the eyes of thoughtful folk, the most important is undoubtedly the publication of Maeterlinck's "Life of the Bee," with its vivid descriptions of apiarian politics, and its suggestive application of human criticism thereto. This great poem has, I say, given to bee-keeping an impetus which any number of mere practical and economic treatises would have failed to afford. As a result, a large number of people, without experience but full of enthusiasm, are eager to commence the practice of this simple branch of farming. For them, in the first place, is this book written. It is believed that in the simplicity of its style and in its systematic arrangement it will compare favourably with all previous works on the practical side of bee-keeping—excellent though a few of these are. The beginner is likely to obtain great assistance from the detailed description of the various appliances used, and many will be glad to be provided with working drawings of that best of all hives, the W.B.C., for which bee-keepers can never repay Mr W. B. Carr, its inventor. The illustrations are very numerous and will be of interest to experienced bee-keepers as well as to the novice. It is believed that the photographs reproduced in this book are beyond comparison with anything previously attempted in the way of illustrations of bee-life.

The Editor desires to express his thanks for assistance rendered by the following gentlemen and firms, either in the form of advice or of the granting of permission to illustrate certain appliances manufactured by them :— The editors of the *British Bee Journal*, which should be supported, read and studied by everyone who possesses even a single hive; Mr F. W. L. Sladen, of Ripple Court, near Dover; Mr W. P. Meadows, of Syston, near Leicester; Mr E. H. Taylor, of Welwyn, Herts; Messrs Abbott Bros., of Southall, near London; and Mr John H. Howard, of Holme, near Peterborough.

SECTION I

BEES—THE ARRANGEMENT OF THE APIARY—BEE PASTURAGE

CHAPTER I

INTRODUCTORY

THAT the keeping of bees has a distinct, well-nigh undefinable fascination, peculiarly its own, no ardent beekeeper will deny. "The genial Baron von Ehrenfels (to quote Dzierzon), who has called beekeeping the 'Poetry of Agriculture,' could not have expressed more beautifully the charm which beekeeping possesses."

Beekeeping is also a centre, a starting point, from which many delightful roads radiate; and creates a stimulus for increased knowledge, not only of the habits and life history of the little creatures who so delightfully minister both to our pleasure and profit, but also of the truly marvellous part they (with other allied insects) play in the economy of nature.

Probably no other pursuit is calculated to bring one into intimate touch with nature at so many points as is beekeeping.

To the gardener, the fruit-grower especially, bees are an absolute necessity.

Flowers are the sexual organs of plants, the male element being represented by the stamens, the pollen grains of which, constituting the fertilising material, require to be conveyed to the pistil or female element, before fruit can be produced.

Every gardener knows the value of shaking, say a vine or tomato plant, when in flower in order to "set" the fruit, and doubtless both wind and rain to a certain limited extent act in a similar manner, yet the generality of fruit blossoms, although hermaphrodite, are as a rule incapable of self-fertilisation. Some plants such as the cucumber, vegetable marrow, etc., produce distinctive male and female blossoms, and apart from troublesome artificial means such flowers can only be fertilised by insect agency.

In the case of such plants as the raspberry, blackberry, strawberry and allied fruits, each ovule or seed requires a separate fertilisation, and it has been computed that a perfect strawberry represents from one to three hundred fertilisations.

Bees when foraging for the sweets secreted by the nectaries carry pollen from flower to flower, thus ensuring the necessary fertilisation of the blooms, totally unconscious of the important part they are playing in the economy of nature.

Bees use pollen in considerable quantity in the springtime for the preparation of brood food, which pollen they convey to their hives in the form of little pellets, snugly stored in the pollen-baskets with which their hind legs are furnished; and although pollen varies in colour according to the source from which it is obtained, it is interesting to notice that the two hind legs of a pollen-laden bee are invariably of the same colour, showing that a bee during one journey gathers pollen solely from blossoms of one species.

Whilst the bees are thus contributing so much to the success of our fruit-crops, they are at the same time harvesting another store of riches in the shape of surplus honey, the extent of which will vary considerably according to the season and the skill of the beekeeper.

To be a successful beekeeper requires no extraordinary amount of specialised knowledge. Anyone contemplating embarking in the pursuit should possess a fair amount of nerve, and should be neat, orderly, and methodical in habit.

To do the *right thing* at the *right time* spells success in beekeeping.

Financially, beekeeping will be found to yield a larger return for capital invested than any other rural industry, and beekeeping is as yet apparently by no means overdone, judging by the amount of honey annually imported from abroad.

Those who contemplate embarking in this pursuit are strongly recommended to do so on a limited scale only, until sufficient experience has been gained. Having proved their aptitude as beekeepers, the apiary may be increased. On the other hand a start with one stock only is equally to be deprecated, as having no standard of comparison; and the embryo beemaster will always be either in grave doubt as to the well-being of his stock, or on the other hand he will be unduly optimistic.

Again, having once overcome his natural fears (a desirable consummation usually speedily attained) the temptation to be always opening the hive and examining the bees is almost irresistible, yet such overmanipulation seriously interferes with the well-being of the colony; but having two or say three stocks, one of them can be selected for experimental manipulations, the remaining colony or colonies being left severely alone, and only interfered with when absolute necessity compels.

CHAPTER II

A FAIR knowledge of the habits and natural history of the bee is absolutely necessary to successful management.

During the summer months a colony of bees will be found to consist of three kinds, viz. :—

1. A Queen,
2. Drones,
3. Workers.

The Queen.—The queen, only one of which is tolerated in a colony, is distinguishable from the other inhabitants of the hive by her greater size. The body is long and tapering, and is only half covered by the wings. The head is rounder than that of the worker, and the abdomen somewhat lighter in colour. The sting is curved, but is used only when in combat with a rival. Consequently queens may be freely handled without fear of consequent stings. Her legs are longer than those of the worker bee, and as she gathers no pollen are devoid of either brushes or baskets. Her special function is to lay eggs from which are raised all the other inhabitants of the hive.

A really good queen will lay as many as three thousand eggs per day of twenty-four hours, and it has been computed that a queen will during her life-time lay eggs equal to 110 times her own weight.

A queen will usually live from four to five years, but after her first complete season her powers will usually begin to wane more or less, and stocks headed by queens more than two years old are almost certain to swarm. Therefore one of the first conditions essential to success in beekeeping and the prevention of swarming is to always take care that colonies are headed by young and vigorous queens.

During the winter the queen ceases to lay, but resumes egg laying in the early spring, gradually increasing the number of eggs as the weather grows warmer until the maximum amount is reached, when after the honey harvest is over the eggs decrease in number day by day until ovipositing ceases for the season.

Queens never leave the hive excepting for the purpose of mating with a drone, or when heading a swarm.

The Drones.—The drones are the male bees, and are recognisable by their ungainly lumbering motions. In size they are intermediate between the queen and the worker. They are stingless, and their primary function is the fertilisation of the queen bee. Therefore the drone is an essential factor in the perpetuating of the species. This fact is so well known to the bees that no colony will swarm unless drones be present to ensure that the future queen shall be impregnated. Towards the close of the season the drones are ignominiously driven from the hive by the workers.

Queenless stocks will tolerate drones at a time when colonies headed by a fertile queen have cast out all drones. This unseasonable toleration is due to a lingering hope on the part of the bees of being able to raise a new queen, in which case drones would be required for mating purposes.

Their note when on the wing is characteristic, hence their name. They gather no honey.

The Workers.—These constitute the main population of the hive, and during the honey flow may number from forty to fifty thousand and upwards. Physiologically, they are undeveloped females so far as the ovaries are concerned. The brain, however, is much larger than the brain of either the queen or drone, and the glandular system is most highly developed. With the exception of reproducing their species, the whole work of the hive is performed by them. Upon them devolves the secretion of the wax, the building of the combs, the gathering of honey, pollen and propolis, this latter body being generally gummy and resinous substances exuded by the leaf buds and the bark of various trees, and used by the bees as a cement. The workers also prepare food for and nurse the larvæ. They constitute the ruling spirit of the hive, regulating entirely its internal economy.

The life of the worker bee during the summer seldom lasts longer than from four to six weeks. During the winter they hybernate, and in this condition may live well into the following spring.

They are also furnished with stings, which they use in defence of their stores.

DRONE.

QUEEN

WORKER

(From a photograph by John C. Douglas)

CHAPTER III

THE DOMESTIC ECONOMY OF THE HIVE

AFTER issuing from the parent hive, a natural swarm of bees, if left to itself, very soon finds out some suitable abode, and having taken possession, forthwith sets to work to furnish it. Previous to swarming, the workers each imbibe honey freely. (In passing, it is of practical interest to note that, when gorged with honey, bees are but seldom inclined to sting, hence as a general rule swarms may be freely handled.) Their object in thus partaking freely of honey is not so much the furnishing of themselves with a supply of food as that of laying in a store of raw material for comb-building, for wax is a natural secretion of the bee, and exudes in the form of scales in what are called the wax-pockets beneath the abdomen of the worker. For every pound of wax produced the bees require to consume from 10 to 20 lbs. of honey.

Comb-building commences at once, and is carried on with astonishing rapidity, so as to provide the waiting queen with suitable receptacles in which to deposit her eggs. These she proceeds to lay as soon as sufficient comb is produced. The cells are of three kinds, those in which worker eggs are laid and honey stored constituting by far the greater portions of the combs. These measure five to the inch in width, and in thickness about seven-eighths of an inch, the latter measurement comprising two cells back to back with one common base. When used for storing honey in the

7

upper portions of the comb, the bees frequently elongate these cells considerably.

Drone cells, *i.e.* cells in which drones are reared, are larger than the worker cells, and measure four to the inch. These cells are generally built in the lower portion of the combs.

The third variety of cell, in which the queens are reared, called queen cells, are comparatively few, and are only built when the bees are queenless, or in preparation for swarming. After serving the purpose for which they were built they are destroyed, and the wax utilised in other ways.

In depositing her eggs, the queen makes careful examination of the cell, and if found suitable to her requirements she turns round, and inserts her abdomen, the length of which enables her to deposit the egg at the bottom of the cell. These movements are repeated with monotonous regularity, the queen being fed continuously by the workers.

The eggs deposited in the cells are of two kinds only, viz. worker eggs and drone eggs, and the queen invariably deposits each in its proper cell, the worker eggs in worker comb and the drone eggs in drone comb.

After laying the egg, the queen has no further interest in her progeny ; the clustering bees maintain the temperature requisite to hatch out the eggs, which takes place in from three to four days, the result being a tiny pearly-white grub, which is at once fed by the young worker bees with what has been termed " chyle food " for (in the case of a worker) a period of three days, then for a further period of two days the larva is fed on a modified chyle food. This is usually termed weaning. During all this time it casts its skin several times, and grows rapidly in size, accommodating itself to its limited surrounding by coiling up in the bottom of the cell somewhat like the letter C. The cell is next

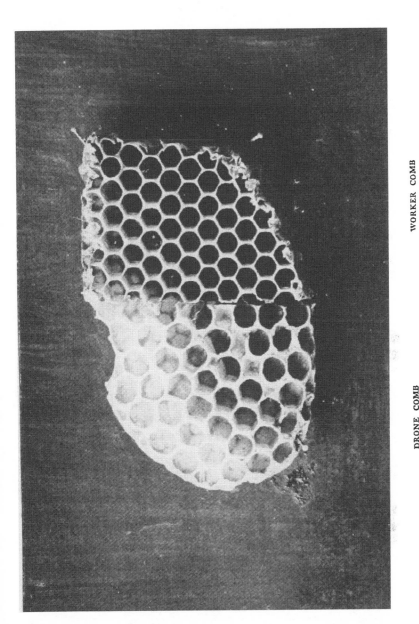

DRONE COMB   WORKER COMB

(From a photograph by John C. Douglas)

sealed over with a convex porous capping of wax and pollen, after which the imprisoned larvæ completes its last moult, spins a cocoon, and is now variously termed chrysalis, pupa, or nymph. Finally, in about twenty - one days from the laying of the egg, the insect emerges from its little prison, a perfect bee, liberating itself by biting its way through the capping of the cell. This last process can often be observed when removing combs from " driven " skeps, and can then be conveniently watched. In appearance the young bee is much lighter in colour than her elder sisters. The day following her birth she commences her work as a nurse bee, feeding the young larvæ with chyle food as previously described, in which occupation she continues for about a fortnight, after which she flies abroad as a gatherer of nectar or pollen.

Drone eggs take rather longer to complete their various metamorphoses, the perfect insect emerging on the twenty-fourth day after the egg is laid. The sealed drone brood is to be distinguished from sealed worker brood, firstly, by its being deposited in the larger drone cells ; secondly, by the greater convexity of the cappings. Both are distinguished from sealed honey by the cappings being much darker in colour, owing to the admixture with pollen. Honey, as a rule, is only stored in the upper portions of the comb.

It now remains to consider the manner in which the queen bee originates. It was stated in the earlier portion of this chapter that the queen laid eggs of two kinds only, viz., those destined to produce workers and those for the production of drones. If from any cause the bees require a new queen they construct queen cells. These are of a special shape and size, and are usually built upon the edges of the combs. In appearance they resemble acorn cups, and hang downwards. The larva produced from an ordinary worker egg laid in such a cell is fed by the nurse bees with food of great richness and in greatest abundance (termed by Huber "royal jelly") for the full period of its larval existence, and is not weaned, as is the case of larvæ intended to produce workers. The result of this liberal diet is fully to develop the ovaries, thereby ultimately producing a bee capable of reproducing its species. The fully matured virgin queen emerges from the cell in fifteen or sixteen days from the time the egg was laid.

A glance at the illustrations (pages 10-90) taken from life will show the appearance of the sealed queen cell. Five days after birth or thereabouts the young virgin queen sallies forth for the purpose of being fecundated by mating with a drone. Fertilisation must take place during flight, and is probably Nature's safeguard against inbreeding. The act costs the drone its life, and the queen returns to the hive with portions of the male organs attached, of which she eventually rids herself. The ovaries themselves are not actually fertilised, but the queen receives and stores the spermatozoa of the drone in a special receptacle provided for that purpose, and she can fertilise or not as she wills the eggs that she lays. The bees produced from fertilised eggs eventually become either workers or queens at the will of the workers themselves. The unfertilised eggs produce drones.

From this it will be apparent that queens which have never mated with a drone are yet capable of laying eggs, but these eggs invariably produce drones. Still further, the drone with which the queen mated has no influence upon her male progeny. This is demonstrable by cross fertilising—say, a native queen with an Italian drone. The resultant worker progeny are all hybrids, whilst the drones retain all the characteristics of the pure native. This must be taken as the general rule; only there are exceptions.

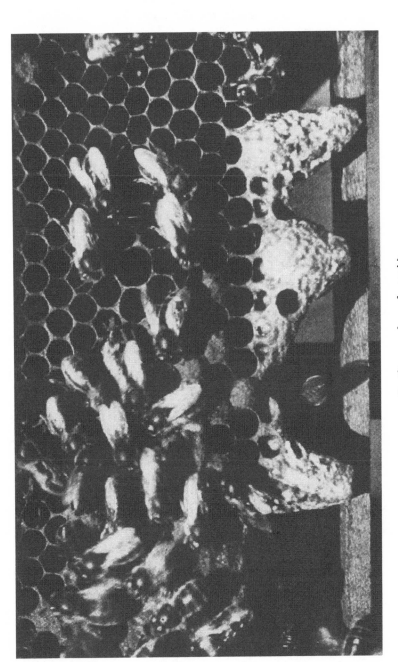

QUEEN CELLS (built on edge of comb)
(From a photograph by John C. Douglas)

The influence of the Italian drone, as regards the drones, would only manifest itself in the offspring of a queen raised from the worker brood of the original crossed queen.

Paradoxical as it may seem, it would appear that the drone has no father, his nearest male parental relative being his grandfather.

As the season advances the queen continues laying eggs, and the young brood hatch out in continually increasing numbers until overcrowding presents a serious problem. The workers solve the difficulty by commencing the construction of queen cells (the presence of drones being assured) and the rearing of young queens. These queen cells after completion become an object of considerable interest to the reigning queen, who tries by all means in her power to tear them open and destroy the young queens therein. This she is prevented from doing by the worker bees, who consistently defend them from all onslaughts. Meanwhile the hive is getting more and more crowded, and its capacity is strained to the utmost, so much so that clusters of bees may be observed hanging around the entrance like miniature bunches of grapes, until one fine day, generally between the hours of 10 A.M. and 4 P.M., the swarm pours out of the hive like a living liquid stream, the old queen amongst them. There seems to be no law governing what bees shall remain behind and what shall join the swarm. Old and young pour out indiscriminately, and pollen-laden bees returning from the fields at the critical moment when the swarm is issuing readily join the excited throng of bees.

The swarm, after circling about in the air for some little time, finally selects an alighting place (such as the bough of a tree) near at hand, the cluster growing larger and larger as the flying bees join their already settled comrades.

How to secure the swarm, and its subsequent treatment, will be dealt with further on (see sec. iii. chap ii.).

The parent stock is now considerably reduced in numbers, and is still queenless ; and until one of the queen cells is hatched, and the young queen successfully mated and laying, but little honey will be gathered ; so that rapid increase of stocks and honey-production cannot go on side by side.

Should the first hatched queen in the parent stock be prevented by the bees from destroying the remaining queen cells, a second swarm or "cast," as it is termed, issues usually about the ninth day following the issue of the "prime" or "top" swarm. Casts ignore both time of day and weather when issuing, and, being headed by a virgin queen, sometimes fly a considerable distance ere clustering. At times further casts are "thrown" off at intervals of one or two days.

As an indication of the probable issue of a cast within one or two days, the queen may sometimes be heard to emit a shrill sound ("piping") as she is foiled in her efforts to destroy the remaining queen cells.

The issue of all secondary swarms is unprofitable, as tending to weaken unduly the parent colony.

VARIETIES OF BEES

In addition to the ordinary brown or English bee, other varieties have at various times been introduced and cultivated, with a considerable amount of success.

Chief amongst these is the Italian or Ligurian bee, first introduced into this country in 1859 by the late Mr T. W. Woodbury, of Exeter.

The Ligurian Bee is mainly characterised by three yellow bands around the abdomen. It is claimed for this variety that it will work both earlier and later, and is of a milder disposition than the common brown bee. It is very energetic, and is said to be able to gather honey from flowers, the nectaries of which the common brown bee cannot reach.

This variety is a great favourite with American bee-keepers.

The Cyprian Bee is somewhat smaller than the Italian, and is a native of the island of Cyprus. It has an unenviable reputation as being exceedingly vicious. On the other hand, it is claimed that this unfortunate trait is entirely absent in pure-bred Cyprians; the bees hitherto looked upon as Cyprians being really a cross between Cyprian and Syrian bees.

In colour it much resembles the Ligurian bee, but is brighter.

The Carniolan Bee, from Carniola in Austria, is remarkably good-tempered, and has been aptly named the "ladies'" bee. In appearance it somewhat resembles our own native brown bee, but the abdominal rings are lighter in colour. The comb honey is beautifully white, and in appearance is unsurpassed. The great fault with Carniolans is their tendency to excessive swarming.

Hybrids.—These are cross-bred bees, the parents being of different races. A strain of bees is undoubtedly improved by the introduction of fresh blood, but the great danger of indiscriminate cross-breeding lies in the fact that the resultant race is likely to prove anything but amiable.

In the hands of a skilled breeder, hybrids may be raised just as gentle in disposition as any race of pure-bred bees (see Queen Rearing).

For all round excellence, however, it will be found hard to beat the common English brown or black bee, and the beginner in bee-keeping is advised to make his initial essay with this variety.

CHAPTER IV

THE APIARY AND ITS ARRANGEMENT

THE amateur beekeeper has usually but little choice as regards the situation of his proposed apiary. Whilst south and south-east are the most desirable directions in which the hive entrances should face, there is no reason why they should not face in almost any direction. With a comparatively large number of hives and a limited space in which to bestow them, it is far better that the entrances should face in various directions, as tending to lessen the chance of flying bees mistaking other dwellings for their own, besides which a studied uniformity of arrangement tends to destroy that delightful quaintness of appearance which was so characteristic of the old-fashioned bee-garden with its "hackled" skeps. From the point of view of the picturesque, one regrets the passing of the old straw skep, but with due care there is no reason why our modern apiaries should not be almost as pleasing to the eye.

Should the hives be home-made, one can be fitted with legs longer or shorter than another; the roof of one gabled, another with the sloping roof shown in the working drawings. All of which departures from the original design can be carried out without in any way interfering with the interchangeability of the separate parts. By so doing we save dull uniformity in appearance, and at the same time materially assist the bees.

The hives should be so placed that they can be approached from the back for manipulation. See that the floor boards are absolutely level (using a spirit level for the purpose) in order to ensure straight combs and perfectly even sections. In comb building, the bees suspend themselves in a cluster, this cluster gradually descending as the comb progresses towards completion, thus forming a natural plumb-line of bees.

The hives may be placed from three to six feet apart; avoid straight rows of strictly equidistant hives. Place, say, a couple four feet apart, then perhaps an odd one some six or eight feet away, then again a group of three, and so on according to situation and the number of colonies kept.

Shelter from the north and east is desirable when possible, and if due care be taken that the bees are in no way impeded in their flight, fruit trees together with small bush fruits give desirable shade and shelter, besides forming sources of early pollen and honey, and convenient alighting places should swarms unfortunately occur.

The hives should not be situated too near a public thoroughfare ; at the same time avoid a too remote situation. There is a marked difference in the behaviour betwixt bees who are habitually accustomed to people walking amongst their hives, and those who scarce ever see a human being.

Water should be provided, especially in the spring-time. A jar full of water (containing a pinch of salt) inverted over a plate constitutes a ready means of supplying this necessity. Bees, unlike flies, readily drown themselves, and to obviate this, place small pebbles around the inverted jar so as to form convenient alighting places.

CHAPTER V

To the beekeeper the term honey-yielding plant means those plants the nectaries of which are capable of being rifled by the honey-bee. The blooms of many plants secrete nectar in considerable quantity, and are yet quite useless to the beekeeper, the nectaries being so situated as to preclude the possibility of their sweets being appropriated by honey-bees. Plants only encourage the visits of those particular insects by whose agency they are capable of being fertilised, and it is only during the time such insects are abroad that the plants make special efforts to attract their attention. The honeysuckle, for instance, being a moth fertilised plant, only exhales its fullest fragrance at evening time when moths are flying freely.

In order to attract the attention of insects to themselves plants would seem to rely upon either the sense of colour or that of smell; and although many exceptions occur, it will often be found that the sweetest smelling flowers are the most insignificant in appearance, whilst the most gorgeous blooms are devoid of perfume.

Double flowers, being destitute of ovaries, do not require to be fertilised, therefore they secrete no nectar.

The growing of honey-yielding plants in small gardens is of little or no use, the amount of nectar to be derived from such a limited area of forage being quite insignificant. On the other hand much assistance may be rendered to the bees by the cultivation of early pollen-

yielding plants and shrubs, pollen being required by
the bees in abundance during the rearing of brood in
the early spring.

The following list may be taken as indicating the
kinds of plants best suited for this purpose :—

POLLEN-YIELDING PLANTS

Arabis alpina.
Berberis (various sorts).
Blackthorn (Sloe.).
Brooms (various sorts).
Crocus.
Daffodils (and various single
    Narcissi).
Flowering Currant.

Hazel.
Primrose.
Snowdrop.
Wallflower (single kinds
    only).
Willow.

Winter Aconite.

Many of the above give fairly large quantities of honey
in addition to a profusion of pollen, notably the Arabis,
which actually yields more of the former than of the latter.

Where fruit is extensively cultivated, and orchards
abound, the time of blossoming will mark the first real
honey-flow of the season. Unfortunately, however,
it often happens that the bees, owing to the uncertainties
of the weather, are unable to take full advantage of
their opportunities at this season.

Following the fruit blossom, hawthorn, and similar
early sources of honey, a period of scarcity frequently
occurs, and it is particularly at this period (to anticipate
somewhat) that the bees require the most careful attention as regards artificial feeding.

The source of the main honey crop will depend upon
the bee forage available in the particular district in which
the apiary is situated.

The following list is intended merely to indicate the
principal sources of honey in this country :—

B

Clovers (various).

Ladies' bedstraw (Galium verum).

Charlock.

Mustard.

Bird's-foot trefoil (Lotus corniculatus).

Wild Thyme.

Horse Chestnut.

Lime

Blackberry.

Lucerne.

Rape.

Sycamore.

Heather.

LING
(*Calluna Vulgaris*)

BELL HEATHER
(*Erica Cinerea*)

Of heather there are three kinds met with in the wild state in Britain. The honey gathered from the true heather, *Calluna vulgaris* or Ling, will not extract owing to its consistency; the combs require to be pressed in a suitable honey press in order to obtain the honey. The Ling secretes nectar more abundantly than either of the other two native heaths.

Next in order as regards honey value stands the *Erica Cinerea* or bell heather (the specific name *Cinerea* having reference to the ashy greyness of the stems), found in great profusion in the West, as also in Scotland, Ireland and Wales.

CROSS-LEAVED HEATH
(*Erica tetralix*)

The honey from this can be readily extracted by means of the centrifugal extractor, if the operation be not delayed too long after the honey has been sealed over.

The third variety, *Erica tetralix*, cross-leaved heath, cannot be considered as a honey plant. The plant is smaller and the flowers larger than either of the foregoing varieties, and is usually found growing in moist situations. Parkinson calls this plant the low Dutch heath.

As to the amount of honey secreted by the various honey-yielding plants, the quantity will vary considerably according to the season ; some plants yielding large quantities of nectar one year, yielding very little the next.

Simmins states that Clover, Sainfoin, etc., produce ten pounds of honey per acre each fine day, and that the flow should last about fourteen days.

Bees will forage within a two mile radius of their hives.

SECTION II

ON BEE APPLIANCES AND THEIR USES

CHAPTER I

BEE APPLIANCES—THE HIVE

HAVING made up one's mind to keep bees it is absolutely essential that all appliances necessary for their well-being should be procured and be in readiness before the bees themselves are acquired.

The following is a list of the articles necessary:—

1. The hive.
2. Frames.
3. W.B.C. metal ends.
4. Foundation.
5. Wire for fixing foundation.
6. Woiblet spur embedder.
7. Quilts.
8. Smoker.
9. Veil and gauntlets.
10. Slow feeder.

The above appliances suffice for the comfortable housing and well-being of a stock or swarm until the honey flow commences, when, in order to secure the surplus honey over and above the bees' requirements, further apparatus will be needed, all of which will be found fully described together with all necessary instructions for use under the heading of surplus honey.

The Hive.—The invention of the bar frame (or, to give it its older name, bar-and-frame hive), together with the introduction of comb foundation, has revolutionised beekeeping by laying open to our inspection what had hitherto been a sealed book ; though from time to time attempts had been made to construct hives, the combs of which could be removed for examination and replaced at pleasure.

The earliest step was the introduction of bars, *i.e.* bars of wood placed at regular intervals across the top of the hive, gridiron fashion, so as to afford the bees a starting-point from which to build their combs ; but the combs being extended horizontally and firmly cemented to the walls of the hive prevented their removal.

The next advance was made by one Propokovitsch, a Russian, who constructed a hive in which frames were inserted endways, the frames being supported on cross pieces on the bottom of the hive. The great objection to frames being supported within the hive in this manner lies in the fact that bees firmly cement them to their supports with propolis, thus rendering it well-nigh impossible to remove them.

The next step was to unite the two ideas, viz., the *bar* and the *frame*, and construct a hive the frames of which should be suspended from bars, thus, amongst other advantages, giving the frames the maximum amount of support whilst preventing undue propolisation.

The credit of this idea is due (according to Neighbour) to a Major Munn, an Englishman, who patented in France, in 1841, a movable bar - and - frame hive. Lorenzo Lorraine Langstroth, a congregational minister in America, and one of the pioneers of modern bee-keeping, invented a similar hive in 1852, and Baron von Berlepsch, in Germany, hit upon a similar idea in 1853.

At the outset the beginner is met with the difficulty, what hive to use. The dealers' catalogues only tend to further confusion, so numerous are the different modifications of the bar-frame hive there placed before him.

Undoubtedly the hive best adapted for general use is

that known as "the **W.B.C.**," a hive designed by **Mr W.** Broughton Carr, editor of the *British Bee Journal*

THE " W.B.C." HIVE

Most dealers retail this hive at one guinea. Ten standard frames, shallow frame super with frames, and outer lift, one rack of sections, and queen excluder, together with a second outer lift, are usually included at the before-named price.

It is well within the powers of an amateur carpenter

to construct these hives for himself, and at about one-third of the catalogue price.

The following description of this hive, together with the drawings appended, will be found sufficiently detailed to enable anyone with care to construct his own hives :—

The Floor Board is built up on a couple of joists AA, and rests upon a separate four-legged stand. (This loose stand constitutes one of the essential features of the " W.B.C." hive.) It will be noticed that the portions C and D are raised a step above the actual floor level. This is done with the idea of keeping the floor dry during driving rain.

The Brood Chamber.—Resting upon the floor board, but not attached to it, is the brood chamber. This is so called because in it, when fitted with frames containing the embedded wax sheets called foundation, the queen lays her eggs and the young bees are reared. When resting on the floor board the strip E (which is loose) forms a bridge across C C of the floor, leaving an entrance $\frac{3}{8}$ of an inch high by $14\frac{1}{2}$ inches wide.

The overhanging lugs of the frames (see frame) rest on the metal strips D, and run at right angles to the entrance. By arranging the frames so, instead of parallel with the entrance, the bees have equally ready access to any comb in the hive.

Great care should be taken that the top bars of the frames when placed in position are exactly flush with the top of the brood chamber, and also to see that there is exactly $\frac{1}{4}$ of an inch clearance (neither more nor less) between the sides of the frames and the sides, or rather the back and front, of the brood chamber. It has been found that bees leave a $\frac{1}{4}$ inch passage-way untouched ; a wider space they would fill with brace combs, a smaller they would propolise.

Brood Lift.—Likewise resting on the floor board, but not attached to it, rests the brood lift, upon the front

part of which a porch is built.   It surrounds the
brood chamber, giving an air space of about 1½ inches
all round.   This tends to make the hive less sensitive
to sudden changes of temperature.   The brood lift
is held in position by means of rabbetted fillets on
three sides which overlap the floor board.   A groove
is made beneath the porch in which slides the
shutter (two strips of wood), thus forming an ad-
justable entrance.

The Super.—This is in all
respects similar to the brood
chamber, only shallower, and
is used to contain the shallow
frames used for extracting.

When working for comb
honey, a section rack takes the
place of this shallow super.

Shallow Lift. — Whatever
form of super be used, whether
for shallow frames or sections,
it is enclosed within a shallow
lift of exactly the same dimensions as the brood lift excepting
as regards depth.

BRASS CONE

At least a couple of these lifts should be provided for
each hive, and one or two extra ones may prove useful
in an emergency.

The Roof may be made in a variety of shapes according
to individual fancy.   Perhaps the simplest is the form
figured in the working drawing.   The framework supporting the roof proper is made of such a size as to
slide over the sides of the hive, fillets being thus dispensed with.   Ventilation is provided by means of a 1½
inch hole in the front.   To afford stray bees a means
of escape, whilst denying admittance to those on plunder
intent, this hole is fitted with two perforated brass

cones, one nailed on the outside, and the other on the inside, both pointing outwards.

Note.—This form of roof differs from Mr Carr's design, as shown on page 23; and is suggested, not as an improvement, but as being somewhat easier of construction by an amateur carpenter.

CHAPTER II

THE HOME-MADE "W.B.C." HIVE

THE wood used for hive-making need only be planed on one side, but should be thoroughly well seasoned, the best for the purpose being yellow pine that has been stored for at least two years. More than one appliance dealer will supply properly seasoned wood cut into suitable lengths should any difficulty be experienced in obtaining this locally.

Most of the nailing together may be done with $1\frac{1}{2}$-inch oval wire brads; for the rabbetted fillets use 1-inch oval wire brads. In nailing the floor-board to the joists much stronger nails should be used.

The life of the legs may be indefinitely prolonged if a hole an inch in diameter and an inch in depth be bored in an upward direction. These holes should be filled with crude creosote, and the legs allowed to stand in an inverted position until the preservative be all absorbed.

The following tables give the measurements of the various parts as shown in the working drawing :—

The Floor-Board.

	No. of Pieces.	Length.	Width.	Thickness.
A	2	$28\frac{1}{2}$ inches.	$2\frac{1}{2}$ inches.	$1\frac{1}{2}$ inches.
B	1	19 ,,	23 ,,	$\frac{1}{2}$,,
C	2	$20\frac{1}{2}$,,	$2\frac{1}{4}$,,	$\frac{3}{8}$,,
D	1	17 ,,	$14\frac{1}{2}$,,	$\frac{3}{8}$,,
E	1	19 ,,	6 ,,	$\frac{1}{2}$,,

The Stand.

This is made of such dimensions as to afford a firm support for the floor board (see illustration page 29). The four legs, which should be about 10 inches long, are made from scantling 3 inches by 3 inches, and should be so cut, that when built into the collar or framework of the stand, they splay outwards *from the corners.*

The slightly enlarged drawing A shows how the legs should be cut so as to obtain the necessary splay.

The collar, or frame, is made from ¾ inch by 2½ inches stuff.

Despite appearances, this stand is unequalled for rigidity.

An alternative method of supporting the floor board, by means of four legs screwed direct to the joists A, is shown in the two sectional drawings page 31.

The Brood Lift.

	No. of Pieces.	Length.	Width.	Thickness.
A	2	18 inches.	9 inches.	½ inch.
B	2	25 ,,	9 ,,	½ ,,
C	1	19 ,,	6 ,,	½ ,,
D	1	19 ,,	1 ,,	⅛ ,,
EE, etc.	1	5 ft. 4 ,,	2 ,,	½ ,,

Rabbetted fillet, to be cut into three lengths as required. Corners to be mitred.

F	1	19½ inches.	¼ inch.	¼ inch.

The Shallow Lift.

	No. of Pieces.	Length.	Width.	Thickness.
A	2	18 inches.	6 inches.	½ inch.
B	2	20¾ ,,	6 ,,	½ ,,
CC, etc.	1	7 ft. 1 ,,	2 ,,	½ ,,

To be cut into four lengths as required. Corners to be mitred.

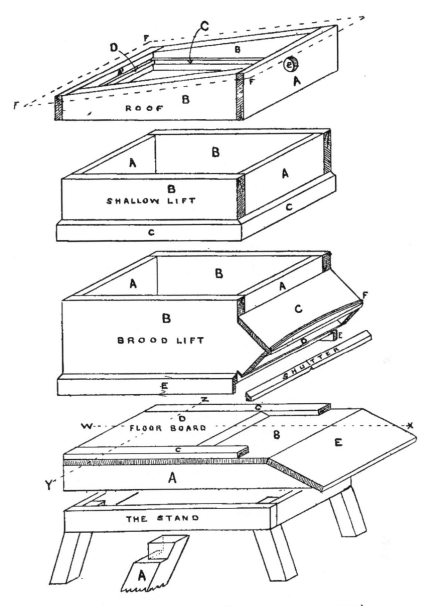

THE HOME-MADE " W.B.C." HIVE (WORKING DRAWINGS)

The Roof.

	No. of Pieces.	Length.	Width.	Thickness.
A	1	$21\frac{1}{8}$ inches.	$5\frac{1}{2}$ inches.	$\frac{7}{8}$ inch.
A1	1	$21\frac{1}{8}$,,	3 ,,	$\frac{7}{8}$,,
B	2	$20\frac{3}{4}$,,	$5\frac{1}{2} \times 3$,,	$\frac{7}{8}$,,
C	2	$20\frac{3}{4}$,,	1 ,,	$\frac{5}{8}$,,
D	2	$18\frac{1}{8}$,,	1 ,,	$\frac{5}{8}$,,
E	Hole, $1\frac{1}{2}$ inches diameter, to be fitted with two brass cones — one inside and one outside.			
F	4	2 ft. $2\frac{1}{2}$ inches.	$6\frac{3}{4}$ inches.	$\frac{3}{4}$ inch.

These pieces constitute the roof covering, and are best made from ordinary grooved and tongued floor boarding. The grooves and tongues should be well painted with thick, white paint previous to nailing together. When nailed on to the framework of the roof the sides will require planing down, so as to make the overlap correspond with the front and back, as indicated by the dotted lines. The roof should be covered with tightly-stretched calico or ticking. In doing this, use copper tacks, and nail along the edges, through a double thickness of the material, so that no frayed edges be left.

Give the whole of the calico one thick coat of good white paint; next nail three strips about 2 inches wide on the outside of the roof in such a position as to cover the joins. (See the illustration—Hiving Bees—where the three strips are clearly visible.) Then give the whole three coats of white paint.

A roof carefully made according to the above directions will prove to be watertight under any circumstances.

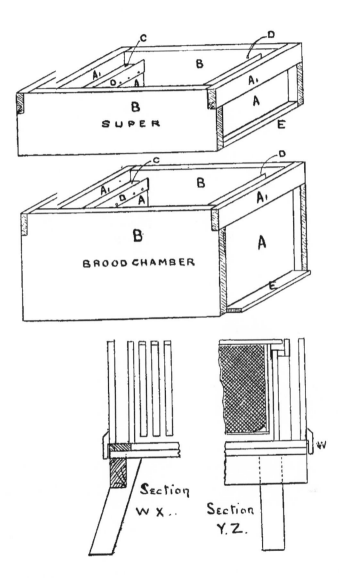

Brood Chamber.

	No. of Pieces.	Length.	Width.	Thickness.
A	2	15 inches.	8¼ inches.	½ inch.
A1	2	16⅛ ,,	1½ ,,	½ ,,
B	2	18 ,,	9 ,,	½ ,,
C	2	15 ,,	¾ ,,	¾ ,,
D	2	15 ,,	1 ,,	{ Stout sheet zinc.
E	1	15 ,,	2 ,,	½ ,,

This piece (E) is not nailed to the brood chamber, but is simply a loose bridge which spans across CC of the floor board, and serves to prevent bees crowding up into the air space. A piece of perforated zinc may be used in place of this, in which case it is better to nail the zinc to the bottom edges of the brood chamber, making it (the zinc) an inch longer than the wooden bridge for the purpose.

In nailing together the four sides AA and BB take particular care that the *inner* surfaces of AA are exactly 14½ inches apart.

A line drawn 1¾ inches from either end of BB should (presuming BB to have been accurately cut) correctly indicates the position of the two pieces AA.

The same applies to the super.

The metal runners DD must be so adjusted as to bring the top bars of the frames exactly flush with the top of the brood chamber. When nailing these into position use an empty frame fitted with the metal ends as a guide.

The same precaution applies to the super.

Shallow Frame Super.

	No. of pieces.	Length.	Width.	Thickness
A	2	15 inches.	5¼ inches.	½ inch.
A1	2	16⅛ ,,	1½ ,,	½ ,,
B	2	18 ,,	6 ,,	½ ,,
C	2	15 ,,	¾ ,,	¾ ,,
D	2	15 ,,	1 ,,	{ Stout sheet zinc.
E	2	15 ,,	1 ,,	½ inch.

The hives, when finished, should be given three coats of good white paint. It is better to have this specially mixed by a reliable oil and colourman as being more economical in the long run. Avoid the use of cheap ready mixed paints.

To paint each hive a different colour, as occasionally advised, with a view of assisting the bees to find their respective domiciles, is to be deprecated. By so doing we at once do away with the first great principle of modern hive construction, viz. the interchangeability of the separate parts.

White is specially selected as the best colour to use, as a white surface reflects the heat of the sun's rays. The combs in a hive painted white will never become so soft as to collapse, whereas such an accident will sometimes happen during the heat of summer in the case of hives painted in darker shades.

CHAPTER III

THE bar frame, which plays such an important part in modern beekeeping, is built up of four separate pieces of wood in the form of a rectangle, the upper part of the frame, or bar proper, from which the frame is suspended, projects $1\frac{1}{2}$ inches on either side of the side bars. The frames are sold in "the flat" for convenience of carriage, and require to be fitted together by the beekeeper.

The British Beekeepers' Association some years ago adopted a standard size for frames, the use of which has now become almost universal throughout this country.

The following are the measurements of the component parts of the standard frame :—

Top bar,	17 inches long,	$\frac{3}{8}$ inch thick,	$\frac{7}{8}$ inch wide.
Side bars,	8 ,,	$\frac{1}{4}$,,	$\frac{7}{8}$,,
Bottom bar,	14 ,,	$\frac{1}{8}$,,	$\frac{7}{8}$,,

When fitted together these form a frame, the outside measurement of which is 14 inches wide by $8\frac{1}{2}$ inches deep, the upper bar projecting $1\frac{1}{2}$ inches on either side. In the hive these projecting lugs rest upon the metal runners, whereby the frame is prevented from touching either the sides or bottom of the hive.

The long upper bar is usually furnished with a saw cut, commencing $1\frac{1}{2}$ inches from either end. This forms a kind of jaw which holds the foundation in position.

Standard frames, as sold by different dealers, vary somewhat in the means adopted to build them up from the separate pieces. Probably the most popular form is the dovetailed frame, which has the great advantage of being easily put together. Another form, invented by Mr E. H. Taylor, Welwyn, Herts, is specially recommended (see illustration). The side bars being $\frac{1}{2}$ an inch thick makes the frame especially strong and rigid, in addition to which the side bars are grooved, giving the foundation additional support. Not being dovetailed, a special frame block is used with these frames whilst nailing them together, so that the completed frame shall be perfectly square.

The construction of this block will be readily understood on referring to the drawing (page 37). Upon a piece of well-seasoned yellow pine, 1 inch thick by 20 inches long and $5\frac{1}{4}$ inches wide, two blocks, AA, of $\frac{7}{8}$ inch wood, are securely screwed; these blocks being 3 inches wide by $4\frac{7}{8}$ inches long. Great care should be taken that the two inner sides of these are exactly 14 inches apart (the width of a standard frame), also that they are strictly parallel with each other. Next in order follow the two pieces, CC, $12\frac{3}{4}$ inches long, 1 inch wide, and $\frac{7}{8}$ of an inch thick. These two pieces are chamfered, as shown in the sectional drawing, and are intended to act as guides for the two sliding pieces, DD, which require to be 6 inches long, and chamfered on the edges to correspond with CC. The two small blocks, BB, are exactly 17 inches apart, each block being set back $1\frac{1}{2}$ inches from the inner face of A A. A hard wood square wedge, tapering from about $1\frac{1}{8}$ inches to $\frac{3}{4}$ of an inch, and about 4 inches long, completes the outfit. E is a square hole cut through the base board upon which the frame block is built.

Any number of frames can be readily nailed together

in a comparatively short time by means of this frame block.

The two sides (grooved faces looking inwards) are placed in the channel between A and D, one on either side; the wedge is next inserted in E, and a tap with a mallet on the wedge makes all secure. Now place the top bar in position by dropping it between BB, and nail together, using one inch oval wire brads. Use one nail only at either end, so that one half of the top bar will come clean away when split in two for the purpose of fixing the sheet of foundation. Next nail on the bottom bar, using two nails at either end, and knock out the wedge when the completed frame can be readily withdrawn from the block.

The dimensions of the frame-making block here given are such as to admit of its use both for standard and shallow frames. These latter differ from the standard frames in depth only, the former being $8\frac{1}{2}$ inches deep, whilst the latter are $5\frac{1}{2}$ inches in depth.

Division or Dummy Boards.—These are used for contracting the hive when less than the full complement of frames is used, and are really curtains of wood hung inside the hive from bars of the same thickness and length as the top bar of the standard frame, viz. $\frac{3}{8}$ of an inch thick and 17 inches long. If the dummy be made of $\frac{1}{2}$-inch board, the width of the upper bar will be $\frac{1}{2}$-inch also. No metal ends are needed, the board being pushed close up in contact with the metal end of the last frame in the hive. When more frames are required, the dummy is pushed backwards and fresh frames are inserted, until the division board is finally withdrawn to make room for the tenth or last frame.

In making division boards (the dimensions of which should be such that they readily slide in and out of the hive), take care to nail cross pieces at each end, so as to prevent warping.

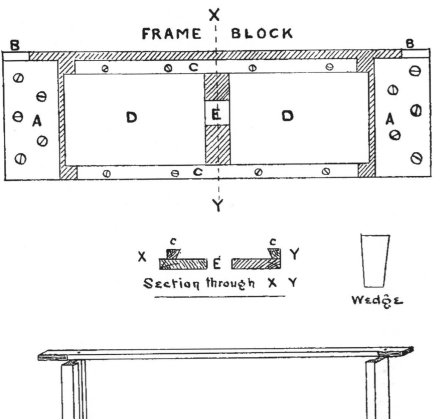

FRAME | BLOCK

X

B   B

A   D   E   D   A

C

Y

X   c   E   c   Y

Section through X Y

Wedge

Standard Frame

Top Bar shewing sawcut

TAYLOR'S FRAME

Metal Ends—Spacing Frames.—The combs in the brood chamber require to be accurately spaced $1\frac{9}{20}$ of an inch apart from centre to centre. This is generally attained

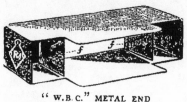

" W.B.C." METAL END

by the use of "metal ends." Those due to the inventive genius of Mr Carr, and known as the "W.B.C." ends, are to be preferred, and, in fact, are almost universally used. They are stamped out of one piece of tin, and in practice are slipped, one at each end, over the top bars of the frames.

For shallow frames many beekeepers use the wide W.B.C. end. The use of this end gives a space of $1\frac{7}{8}$ of an inch, the idea being that with increased space the bees construct deeper cells, thus increasing the honey capacity of each comb, whilst economising foundation and labour in extracting.

It sometimes happens that in using these wide ends the bees will draw out the combs irregularly. To obviate this, use the ordinary - sized ends until the combs are filled with honey and sealed, when, after extracting, the normal ends may be replaced by the wide ones, and the combs be returned to the bees, when regularly drawn-out combs will result.

Comb Foundation.—Foundation, as its name would indicate, consists of sheets of pure beeswax, "honeycombed" all over by having been passed through suitably engraved rollers. Not only are the sheets indented with the bases of the cells, but sufficient wax is left surrounding each cell to enable the bees to draw out the combs to their full extent.

Further, by the use of foundation we are enabled absolutely to control the direction in which the bees shall build their combs, and also to provide that the combs shall be built within the frames, thus making every comb movable and interchangeable ; and lastly, we save the bees an enormous amount of labour by providing them with this raw material, the "foundation" upon which they subsequently build their cells.

That this is no small economy will be readily understood when it is borne in mind that, as previously stated, to produce one pound of wax, bees require to consume on an average something like fifteen pounds of honey.

From the time of Mehring, a German, who in 1857 invented and first used foundation, numerous patents have been issued relating to the manufacture of this commodity, many of them more curious than practical, artifical combs having been made from wood, iron, sheet tin, etc., coated with wax. The foundation in use at the present day is of

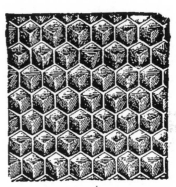

COMB FOUNDATION

two varieties, distinguished by the names of their inventors, A. J. Root and E. B. Weed, and known as "Root" foundation or "Weed" foundation, as the case may be. Both are excellent. The Weed foundation being rolled under much greater pressure than that employed in the Root process, is in consequence much tougher, and is more likely to bear inexperienced handling at the hands of the novice.

Foundation is graded according to the number of sheets required to weigh a pound. Not more than "seven sheets to the pound" should be used for the standard or shallow extracting frames. For comb honey in sections, the thinner the foundation the more pleasing the appearance of the finished sections.

Method of fixing Foundation in the Frames.—Having nailed together a number of frames as previously described, it now remains to fix into each a sheet of foundation. A sheet of foundation will be found to measure about $13\frac{1}{4}$ inches wide by $7\frac{3}{4}$ inches deep, slightly wider than the inside width of the frame but not quite so deep. This is to allow of its stretching as it generally does under the workmanship of the bees. This would give rise to bulging combs were the sheets given the exact size of the frame.

Take one of the frames and break or split off the unnailed half of the top bar. Slide the sheet of foundation into position, pushing it gently downwards in the side grooves, just so far as to leave the upper edge of the sheet flush with the upper surface of the top bar. Or the sheet, by being bent may be sprung into position. Replace the half of the top bar previously broken away, slip on a couple of the metal ends, one at either end, and to make all absolutely secure drive in a couple of nails, one at either end of the broken half of the top bar.

Wiring the Frames.—In order to prevent any possibility of the combs breaking away from the frames it is usual to afford them further support by means of wires, tightly stretched across those parts where the support is the least.

Wiring.—The following will be found a simple and effective method of wiring frames, and does not involve the use of any special appliances.

Four pricker holes are bored, two on either side of the frame opposite to each other, the two lower being about 2 inches up from the bottom bar, and the next two $4\frac{1}{4}$ inches up, or midway between the top and the bottom.

Next take about a yard of tinned iron wire (No. 27 B.W.G.), thread one end through any one hole, draw the wire through and thread through the opposite hole. Then carry the wire on the outside of the side bar, either upwards or downwards as the case may be to the next hole. Repeat the threading through the remaining two holes, and twist tight with the help of pliers. The next operation will be to tighten the wires, and this is best accomplished by pulling the wire at each hole with a pair of round nosed pliers, and then twisting it. The three little loops formed, together with the twisted ends, can then be folded down flush with the frame.

Woiblett Spur Embedder.—The stretched wires now require to be embedded in the foundation. Lay the frame with its sheet of foundation wired side uppermost upon a piece of $\frac{1}{2}$ inch wood, the latter of such a size that it will

WOIBLETT SPUR EMBEDDER

easily fit into the frame. Next take the spur embedder (which is a toothed wheel, furnished with a pulley like groove, mounted in a suitable handle), warm over a lamp flame, and then resting the groove on one of the wires, run the embedder along. At each point of contact the wax will melt and the wire sink into the foundation; the molten wax solidifying holds the wire imprisoned, thus securing the whole.

Foundation should always be so fixed that the points of the hexagons have an upward and downward direction. The flat sides of the hexagons being parallel with the sides of the hive.

Each hive will require ten frames fitted with founda-tion (one and a half pounds of foundation being required for each hive) and wired. When these are placed in the hive, some sort of covering will be required to prevent the bees escaping upwards into the roof, and to keep all snug and warm. Formerly square boards were used for this purpose and were called crown boards.

CHAPTER IV

QUILTS, consisting of squares of unbleached calico, have now, in this country at least, entirely superseded the older crown board. These coverings should be made 16 by 18½ inches, and be provided with a feed hole 3 inches in diameter. This hole should not be in the centre of the quilt but towards the end, so that when the feeder is not in use the hole may be covered by reversing alternate quilts.

Never use less than four calico quilts during the summer time. Felt or flannel quilts should be on hand ready for use during spring and autumn. During the winter the chaff cushion should be used (see sec. iii., chap. xvii., *Preparation for wintering*).

In the early spring, when breeding has begun, a square of American cloth laid over the calico and underneath the felt quilts will materially assist the bees in preserving the necessary temperature for breeding. It is at this season that every means should be taken to conserve to the utmost the internal heat of the hive.

Folded newspapers make excellent quilts in an emergency, but of course should not be placed in direct contact with the frames.

The weak point in the use of quilts is the numerous snug, warm retreats afforded for a variety of insect pests ; but if all material used for quilts be cut to the correct size, the various layers laid neatly, and ordinary care and vigilance exercised, no difficulty of this nature should

43

arise. The careless beekeeper who, when additional covering is indicated, hastily folds up an old damp sack and places it over his bees, together with perhaps a handful or two of straw, is generally rewarded, when he comes to remove this covering, by finding a colony of ants, or maybe a nest of mice, together with earwigs, moths and spiders innumerable.

The quilt next to and in contact with the frames is quickly propolised by the bees. In the autumn when finally overhauling for the winter, replace this propolised

THE SMOKER

quilt with a clean one. The old one had better be burnt at once, and replaced with a new one ready for use the following year.

Propolis may be removed from quilts by exposing them to frost, and then rubbing between the hands; after which wash with hot water and soap.

The Smoker.—This piece of apparatus plays an important part in practical beework, and is so contrived as to place in the hands of the beekeeper a constant smoke supply when manipulating.

The smoker consists of a small wedge-shaped pair of bellows upon which is mounted a cylinder of tinned iron, through which the draught from the bellows circulates, this cylinder terminating in a cone-shaped nozzle.

When required in practice, the nozzle is removed, and loosely twisted brown paper (previously dried if possible) is charged into the cylinder, having first been set on fire at the lower end. The nozzle is next replaced and the fire urged by the bellows, when a plentiful supply of smoke should result, which can be readily directed where required.

Cotton rags or touchwood may be used, but the latter is open to the objection of being liable to emit sparks.

A few experimental trials should be made, and the conditions under which a good fire and plentiful supply of smoke may be obtained, will soon be discovered. Few things are more annoying in practice than to find one's smoker out at a critical moment.

During use, the smoker should be set down nozzle upwards, in which position the fire is kept in by the upward draught induced. When finished with, lay down horizontally, and the fire will soon go out.

The Bingham pattern smoker will generally be found effective, and costs about three shillings and sixpence.

Veil.—As a protection to the face, as also for the greatly increased feeling of confidence gained, a bee-veil should always be worn when handling bees. This may be made of coarse black net, or (with greater comfort to the wearer) of Brussels silk net.

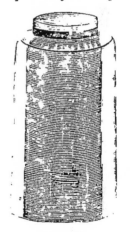

BEE VEIL

A piece of material 1 yard long and 18 to 24 inches wide is amply sufficient. The two ends are to be sewn together so as to form an endless band, the top and bottom should be "hemmed," and elastic inserted into both "hems." The elastic in the upper one should be so adjusted as to fit tightly round the crown of a hat; the lower elastic being much larger, with the two ends projecting like the strings of a bag.

In use the smaller top hem of the veil is drawn over the crown of an ordinary broad-brimmed hat; the bottom hem falls on to the shoulders, and should be tucked inside the coat collar. The loose ends of the elastic are next drawn so as to make the veil fit round the neck, the ends being wrapped round a vest button, and the coat buttoned over all.

Gauntlets should be worn over the coat-sleeves, so as to prevent bees from crawling up the arms. It is also at times advisable to protect the legs by using elastic bands around the bottom of the trousers; ordinary cycle clips may be used instead, or the trousers tucked inside the socks.

Slow or Stimulative Feeder.—This necessary portion of the beekeeper's equipment, the invention of the Rev. C. G. Raynor, is so constructed that the amount of syrup supplied to the bees can be regulated to their requirements. It is used for "stimulative feeding," as it is termed, in spring and in autumn when honey is scarce.

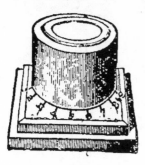

The feeder consists of a wide-mouthed bottle, the mouth of which is closed by a movable metal cap, perforated on the face with, generally, about ten holes, the holes being arranged in a semi-circle. The bottle, after being filled with syrup and capped is inverted upon a wooden stage provided with a central hole, carrying a diaphragm furnished with a semicircular

SLOW FEEDER

slit. If all the holes in the cap coincide with the slit in the diaphragm, the maximum amount of syrup capable of being delivered by the feeder is supplied to the bees.   On the contrary, if the bottle be revolved one half turn, the supply of syrup is shut off, the perforations in the cap being closed by the uncut portion of the diaphragm. The addition of a pointer to the cap, and numbers corresponding to the holes in the wooden stage, provides a ready means of regulating the amount of syrup.

SECTION III

PRACTICAL WORK

CHAPTER I

HOW TO BEGIN

By far the most usual method of commencement is by the purchase or gift of a swarm. The earlier a swarm can be secured the better. The old rhyme, lame and halting as it is, yet speaks true when it says—

> " A swarm of bees in May,
> Is worth a load of hay.
> A swarm of bees in June,
> Is worth a silver spoon.
> A swarm of bees in July,
> Isn't worth a fly."

Having the promise of a swarm, have everything in readiness for its reception some time before its anticipated arrival. The floor board of the hives should be carefully levelled by means of a spirit-level, and the brood chamber should contain about six or seven frames fitted with foundation, and a couple of dummies.

Usually the swarm, if from the immediate neighbourhood, arrives in the skep in which it was taken. This should be placed as received in its natural position in some cool and shady part of the garden until towards evening. In the meantime, in front of the hive that is to become the future home of the swarm, support on bricks or otherwise a board about 18 to 20 inches wide by 2 to 3 feet long, so as to form a continuation of the flight board. This board should slope downwards from the hive in the same manner as the flight board, and the whole should be covered with a small sheet or tablecloth, the upper edge of which should be close to the entrance of the hive, and secured with two or three drawing-pins. The outer case of the hive may be removed entirely, leaving simply the brood chamber with frames and quilts; the former being spaced somewhat more widely than the normal distance indicated by the metal ends. Further, to facilitate the entrance of the bees, the front of the brood chamber should be propped up about an inch with a couple of wedges. Have also in readiness the bottle feeder and a quantity of syrup made according to the receipt for "spring feeding."

The evening of the day on which the swarm issued is the best time to hive it. So, having prepared all things in accordance with the previous instructions, and having donned veil and gauntlets, we proceed to the scene of operations.

Carefully remove from the skep, without inverting, whatever coverings may have been used to secure the bees. Lift up the skep and take a peep inside. The bees will be found clustered in the crown of the skep and lining its sides. Now hold the skep firmly in both hands, mouth downwards, and with a sharp downward throw jerk the bees on to the board in front of the hive, throwing them as close to the entrance as possible. The more violent the jerk (no other word is so expressive) the more successful the operation, and the sudden throw so astonishes and frightens the bees as to subdue all desire to sting, and scarce a bee will even attempt to fly. In a few moments they recover from their astonishment, and those nearest the entrance of the hive soon discover therein a desirable residence, and the intelligence is apparently instantaneously communicated to the whole swarm, for almost at the same moment the entire swarm, from being an apparently aimless crowd, suddenly acquires direction, and is seen to head *en masse* for the hive.

Meanwhile the skep will be found still to contain a "coating" of bees, and should be vigorously slapped (mouth upwards) with the open hands and rapidly swirled round, when the adhering bees will ignominiously roll around in the bottom of the skep like so many peas. A sudden jerky throw will restore all these to their companions, when the skep should be taken right away.

In the upward movement to the hive, the bees move more like a semi-plastic mass than individual units, and frequently clusters form on the edges of the board. These may be coaxed into the right way with the smoker, or be lifted up with a spoon and placed on the board.

After all the bees have entered the hive, which may take from twenty minutes to an hour, or even longer with a heavy swarm, remove the wedges, space the frames to their normal distances, replace the outer cover, fix the feeder, full of syrup, in position, and start gentle feeding (say, two or three holes); lastly, of course, put on the roof. The following morning will usually find the bees flying, feeling apparently quite at home; and if in the course of a day or so they are observed carrying in pollen, then all may be judged to be well and breeding to have commenced.

HIVING BEES

(*From a photograph by John C. Douglas*)

CHAPTER II

TO SECURE OR "TAKE" A SWARM

SHOULD the bees when swarming (see chap. iii., sec. 1, "The Domestic Economy of the Hive") manifest a tendency to fly afield and not to settle readily, they may be induced to do so by throwing water over them from a garden syringe in imitation of rain. Sand, or fine dry garden soil, may be used should neither water nor syringe be at hand.

The bees having finally settled (a few still flying around may be neglected), it only remains to secure them, which should be done as follows : Supposing the bees to have settled on a low-lying branch of a tree or bush, spread a sheet on the ground immediately beneath the cluster, then with one hand hold an inverted straw skep underneath the swarm, and with the other hand give the branch a vigorous shake, when the majority of the bees, with the queen amongst them, will fall into the skep. Reverse the skep, thus bringing it into its natural position, and place on the sheet, propping up one side with a stone or anything handy, when in due time the flying bees will join their companions within.

Should the queen still remain with the few bees adhering to the branch, the bees within the skep will soon discover her absence and will again cluster on the branch, when the operation of shaking, etc., must be repeated. If the particular branch selected by the swarm as an alighting place is of no particular value, it may be gently cut away and the swarm carried to the hive it is to occupy as its future home.

Having secured the swarm in the skep it should next be "run in" to its permanent home in the manner previously described.

Special difficulties will be encountered from time to time. Sometimes a swarm will elect to settle on an old espalier tree nailed firmly to a wall, the branches of which cannot possibly be shaken. In such case the hiving skep may be fixed over the swarm, mouth downwards, of course, and kept in position by resting it on pointed sticks, the prongs of an inverted garden fork, or other suitable means. Cover the whole with an old coat or rug or anything handy, when usually, after a short while, the bees will begin to ascend and take possession of the skep. They may be further urged to do so by fixing a small piece of comb in the skep, together with an occasional upward puff or two of smoke from the smoker.

In the absence of a straw skep, an empty pail, wooden box, or any other handy receptacle may be used into which to shake the bees. Carefully shade the hiving skep after the bees have been shaken in.

Another method of stocking a hive is by purchasing a colony of bees from a dealer of repute, who ought to guarantee that the bees are free from foul brood, that dread disease so feared by all bee-keepers.

A SWARM

(*From a photograph by Charles Reid*)

CHAPTER III

TRANSFERRING

SHOULD the would-be bee-keeper decide to purchase a stock in a skep from some cottager, he should take care to make his purchase in the spring, about the middle of April being the best time, as the excitement set up by removal and transferring provides the requisite amount of stimulation, inducing the queen to commence breeding in real earnest. The bees should have swarmed the previous year, thus ensuring a young and active queen at the head of the colony. Precautions ought also to be taken to ensure that the colony is free from foul brood. Should the colony, the purchase of which is contemplated, be in the immediate neighbourhood, an examination should be made before removal. This should, if the purchaser is a beginner, be made by an experienced bee-keeping friend, or the expert of the county or local Bee-Keeping Association. First watch the bees at work, and observe if pollen is being carried in (a sign of the presence of a queen and that breeding has commenced); next subdue the bees as described under " Driving," and invert the skep. Observe the colour of the combs. If black and rotten, the colony is an old one and an undesirable purchase. Particularly should evidences of foul brood be looked for.

The colony having been pronounced in good health, it should forthwith be removed to the purchaser's apiary, which is best done by first lifting the skep on to a square of cheese cloth, tying the four corners together over the top; string should also be tied around the skep near the mouth, and then, by means of a stick thrust beneath the knots in the crown of the skep, two persons can readily carry it between them without in any way disturbing the bees.

The next process is to transfer the bees from the skep to the frame hive, and this cannot be better done than by the method so often described in the " British Bee Journal."

Procure a square of American cloth, large enough to cover the tops of the frames, in the same manner as do the quilts. In the centre of this cut a hole about 4 inches square. Lay this cloth (glazed side downwards) on the top bars of the frames, then over this place the skep (after subduing the bees with a few puffs of smoke), stop up the entrance to the skep so that the bees are compelled to use the frame hive as an " entrance hall." Cover all space not occupied by the skep as warmly as possible, so as to make the lower hive warm and tempting to the bees. Build up the hive by means of " lifts " sufficiently high so as to admit of the roof covering the whole, and for the time being all is done.

After a few weeks' interval, if the bees are observed to be working vigorously, subdue, and gently lift up the skep and examine the lower hive. If brood be found therein, it may safely be assumed that the queen has left the skep and is laying in the frame hive. This being so, place a queen excluder zinc immediately over the American cloth and replace the skep, so that the bees may return and hatch out any brood that may be present. In three weeks' time, all brood remaining in the skep will have hatched out, and the skep may be removed, together with the excluder and American cloth, and the frames provided with the usual quilts.

Under ordinary circumstances, never purchase a skep in the autumn. There is always the risk of the stock not surviving the winter, and should it perish the purchaser is so much out of pocket.

Apiaries from which stocks are purchased should not be less than a mile away from their new home, otherwise many of the bees, being thoroughly conversant with the neighbourhood, will return to their old homes. Natural swarms invariably note a change of domicile, even though it be only a few feet from the old stand, therefore as regards these the previous precaution does not apply.

CHAPTER IV

It sometimes happens that some particular stock will swarm in spite of all precautions to the contrary. Various methods of dealing with such undesired swarms may be adopted.

1. Return the swarm to the parent stock on the evening of the day the swarm issued, having first cut out all queen cells but one in the parent stock, and also having searched for and destroyed the old queen in the swarm. Until the new queen is hatched, mated, and laying, the population of the hive is necessarily diminishing, with consequent loss of surplus honey.

2. Hive the swarm in a new hive fitted with full sheets of foundation, and placed on the site of the parent stock. This latter may be broken up into nuclei or otherwise dealt with. Supers, if any, should be removed and placed in the new hive. All the *old* bees from the parent stock will fly back to their old location, and thus augment the swarm. This method is to be preferred to the previous one, as entailing less interruption of work. But even then we have to wait three weeks before the new brood hatches.

3. Secure the swarm and run it into a new hive placed exactly on the site of the parent stock, moving the latter a very little to one side, and turning it round, so that the entrance points the opposite way. By this means all the *old* bees will very soon join the swarm. Also transfer supers (should there happen to be any) to the

hive containing the swarm. When the new queen is hatched, mated and laying, destroy the old queen, and on the following evening unite the two lots with the young queen at the head, discarding those frames which contain little or no brood. By this method (due to Mr Simmins) breeding is continuous. Not only so, but we have actually, for a short time, two queens furnishing brood for one hive.

CHAPTER V

WHEN increase of stock is desired, much time is often lost in waiting for the issue of natural swarms. To obviate this is the object of artificial swarming.

Artificial swarming (or more correctly speaking, dividing) should only be attempted—

1. When drones are present.
2. In warm weather when honey is coming in freely.
3. When stocks are strong.
4. During the natural swarming season.

To divide one colony into two.—Between 10 A.M. and 5 P.M. remove from the stock to be divided one frame, containing the queen, bees and brood. Place it in a new hive, filling up the remaining space with frames of foundation. Into the parent hive put a frame of foundation to replace the one taken away, and remove the whole bodily to a new site. The new hive must now occupy the site of the old one. All bees on the wing will join the new hive.

The parent stock, of course, is now queenless. A ripe queen cell may be given or a new queen introduced. (See Introducing Queens.)

To divide two colonies into three,—In this method of procedure one hive provides the brood, and the other the bees.

Hive No. 1.—From this remove four or five frames full of brood, shaking all the bees back again. Replace these with frames of foundation.

Hive No. 2 (which should be a strong colony). Remove to a fresh stand and in its stead place the new hive No. 3, which should contain the frames removed from No. 1.

Hive No. 3.—This now contains brood from No. 1, and all flying bees from No. 2, and will rear a fresh queen, but time will be saved if a new queen be introduced.

A new colony may be formed from a number of others by taking from one hive a frame containing brood bees and queen, which place in a new hive. Fill up this latter with one frame of brood without bees from each of the other colonies. Now remove a strong colony to a fresh site and replace with the new hive.

CHAPTER VI

THE ARTIFICIAL FEEDING OF BEES

BEES may be artificially fed with sugar syrup with the object either of encouraging brood-rearing, or supplying a deficiency of natural stores at the end of the season.

Feeding with the first-named object in view is generally termed "stimulative or slow feeding," for the reason that bees will not raise brood on a diminishing capital; but when capital is supplemented by income, then breeding recommences. This explanation must not be taken too literally. We could not keep bees breeding all the year round by means of slow feeding.

Feeding should be commenced when crocuses and other early spring flowers begin to bloom, and the bees are carrying in pollen in quantity. No definite rule can be laid down as to the precise date to commence, but the best time is about six weeks before the first honey flow may be expected, as it takes about that length of time to build up a colony to full strength.

Syrup for stimulative feeding is made as follows :—

Pure cane sugar . . .	10 lbs.
Water	7 pints
Vinegar	½ oz.
Salt	1 oz.
Napthol beta solution. .	½ oz.

First boil the water, then add the sugar, salt, and vinegar. Boil for one or two minutes, with constant stirring to prevent burning. The syrup having cooled somewhat, add the napthol beta solution with stirring.

The napthol beta solution is added to the syrup as a preventive of "foul brood" (see sec. iv. chap. i., "Diseases of Bees"), and is prepared as follows :—

> [1] Napthol beta crystals . . . 1 oz.
> Rectified spirit . . . . 7 ozs.

Shake well until dissolved. Good methylated spirit may be substituted for the rectified spirit. Half a fluid ounce of this solution is sufficient to medicate the syrup produced from ten pounds of sugar.

The syrup is given to the bees by means of the bottle feeder, described on page 46. Give the syrup slightly warm, about a quarter of a pint each night, turning the indicator so that only two or three holes are exposed. Feeding should be carried on during the night only, as considerable excitement is set up otherwise, and the bees are tempted to fly abroad.

Before commencing to feed in the early spring, remove the outside combs not covered with bees, placing them behind the division boards, which latter should be closed up, but do not unnecessarily expose the bees. Simply unroll the quilts from either end, removing the empty frames, until the first seam of bees is seen. Immediately over the calico quilts place one made of American cloth (with feed-hole corresponding to those in the quilts beneath), glazed side downwards, for at this season humidity is a distinct advantage, as tending to conserve the natural heat of the brood nest to the utmost. Moisture may even at times be seen to trickle out at the entrance. This is a good sign, showing that the bees are active and that breeding is progressing apace.

The feeder, when in use, should be wrapped up snugly,

---

[1] Napthol beta may be obtained from almost all appliance dealers, or from the *British Bee Journal*, 17 King William Street, Strand, London, W.C., at one shilling per packet post free, together with directions for use.

so as to prevent the syrup from becoming chilled. Thick felt quilts are very useful, and should be placed over the American cloth. These should have square holes cut in them so as to fit snugly around the stage of the feeder.

Replace the empty combs, previously removed, at intervals according to the bees' requirements, selecting the middle of a bright, warm day for the purpose. If pollen-yielding plants are scarce, artificial pollen in the form of pea-flour will be readily accepted by the bees. In a sunny corner of the garden place a box or old skep containing shavings or chopped straw ; over this liberally sprinkle the pea-flour, and in a short time the bees will be found availing themselves of this plentiful supply.

Stimulative feeding should also be practised at the close of the honey season, so as to encourage the breeding of young bees. This is an important aid to successful wintering. Also, whenever there is a cessation of the honey flow—as, for instance, between the fruit blossom and white clover bloom—feeding should be carried on, so that the stocks receive no check.

Never feed when supers are on, as there is a great probability that some of the syrup will be stored in the supers.

Feeding with the object of supplying stores for winter consumption is termed " rapid feeding," the syrup being given to the bees in profusion. The syrup for rapid feeding in the autumn should be more concentrated than that used for stimulative purposes, and is of the following composition :—

Pure cane sugar	10 lbs.
Water	5 pints.
Vinegar	$\frac{1}{2}$ ounce.
Salt	1 ounce.
Napthol beta solution	$\frac{1}{2}$ ounce.

Prepare as in the manner before described.

This syrup is given by means of the rapid feeder, several forms of which have been devised. The Canadian rapid feeder works admirably. This consists of a tin-lined wood tank enclosed in an outer wood case, with an interspace of about three-quarters of an inch between the tank and the outer case. This is placed over the brood nest in the same way as a super or section crate. The syrup is placed in the tank (which will hold at least 20 lbs.). On the syrup floats a grating made of narrow laths of wood. A lid covers the whole, and

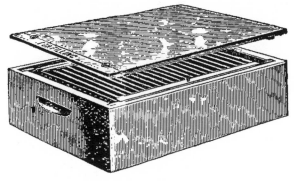

RAPID FEEDER

means are provided for pouring in the syrup without affording an exit for the bees.

The bees crowd up the interspace between the tank and the outer body, and, alighting upon the raft, commence to transfer the syrup to the combs in the brood nest below. The rapid manner in which the bees succeed in taking down this syrup is almost incredible.

Feeding should be continued (at night only, for reasons before described) until all stocks contain about 20 lbs. to 30 lbs. of sealed stores. All unsealed stores should be extracted before wintering. Thirty pounds of sealed stores represents roughly eight standard frames, each three parts filled.

All feeding should be finished, and the bees packed up for the winter, by the middle of September or the first of October at the latest. (See chap. xvii., "Preparations for the Winter").

Should supplies from any cause run short during the winter, candy is the only possible food. To prepare this take :—

Pure cane sugar	.	.	.	6 lbs.
Water .	.	.	.	.	1 pint.
Cream of tartar	.	.	.	1 teaspoonful.

Dissolve in an enamelled pan over a slow fire. When solution is complete raise the whole to boiling point, briskly stirring so as to avoid burning. Allow it to boil for one minute, withdraw from the fire, and drop a small quantity on to a cold plate. If at all sticky when touched, boil again for another minute, until a drop tested as above does not stick to the fingers. At once remove it from the fire, add half a tablespoonful of the napthol beta solution, plunge the pan into a larger one containing cold water, and stir briskly until the mixture turns white and begins to stiffen. Pour into square dripping tins lined with paper and allow to cool. Cakes weighing about two pounds each are a convenient size.

Well made candy, although perfectly stiff, should be smooth in grain, and should be readily scraped into a soft buttery consistency.

Most dealers in bee-keeping appliances make candy, and it is perhaps advisable to purchase an odd pound as a criterion when first candy-making is essayed.

Should either syrup or candy unfortunately be burnt during manufacture, on no account give it to the bees; burnt sugar being most injurious. Neither should beet sugar be used in the preparation of bee foods for a similar reason.

Candy made as just described, but with the addition

of one pound of pea flour stirred into it as soon as it is taken from off the fire, is very useful in the early spring for stimulative feeding. This preparation usually goes by the name of " flour candy."

Candy is given to the bees by simply laying the cakes flat on the top of the frames, immediately over the cluster; covering all up warm and snug with the quilts.

CHAPTER VII

SUBDUING AND THE MANIPULATION OF BEES

To the novice the sting of the bee looms large, and inspires more or less dread according to individual temperament, and no doubt many are deterred from keeping bees simply from the perfectly natural fear of being stung ; this fear in many instances, being out of all proportion to the pain inflicted by an occasional sting.

Under certain conditions, and under certain circumstances, bees are far less disposed to sting than at other times, and the beekeeper who seizes upon these circumstances or conditions when they arise, or who artificially creates them, may safely keep his bees, and examine his hives with but little risk of being stung.

Bees are usually quiet :—

1. In the middle of the day during the honey flow.

2. Immediately after swarming, when gorged with honey.

Bees are disposed to be irritable :—

1. When no honey is coming in, and all stores are sealed over.

2. When queenless.

3. When robbing has started in earnest.

4. During dull or rainy weather when confined to their hives.

When a bee is gorged with honey it is seldom inclined to sting. This may possibly be due to a comfortable after dinner sort of feeling, but is more probably due to fear ; for when really frightened the bee's first impulse

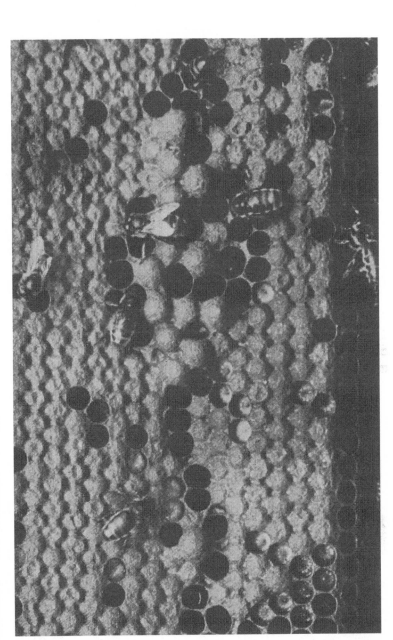

SEALED WORKER AND DRONE BROOD, WITH UNSEALED LARVÆ.

(From a photograph by John C. Douglas)

is to gorge. We have already seen, that previous to swarming, bees imbibe three or four days' supply of honey, and in consequence are at such times usually quite docile. This is to enable them to commence comb building immediately upon entering their new dwelling.

The blowing of smoke, or carbolic vapour, into a hive creates instant alarm, and the bee's idea is that its present dwelling is threatened, therefore it behoves each individual bee to provide for the establishment of another dwelling by taking with it as much raw material for comb building as is possible under the circumstances, and the bees become, as at the time of swarming, gorged with honey. It should be understood that the smoke in no sense stupifies the bees ; it is simply intended, by alarming them, to induce them to imbibe honey freely.

When desirous of opening a hive in order to make an examination, first see that the smoker is in good working order, and burning freely, and after having donned veil and gauntlets, approach the hive to be opened. Gently blow one or two puffs of smoke in at the entrance to drive back the guards, wait a few moments, then lift off the roof, taking great care not to jar the hive in any way. Now, holding the smoker in the right hand, with the left gently raise one corner of the quilt, directing smoke from the smoker on to the top of the frames as they become exposed to view, until the quilt has been entirely removed. The quilt will be found to be more or less firmly glued down to the frames with propolis, but by proceeding slowly, and at the same time firmly, it is possible to remove it without jarring the frames.

The next operation is to take out the frames one by one for examination. To do this, stand the smoker somewhere handy, nozzle upwards. Draw the "dummy" aside, away from the first frame. If firmly propolised use a small screwdriver as a lever underneath the lugs. Having moved the dummy, draw back the first frame until it is midway between the dummy and the second frame, then taking a firm hold of each lug gently lift it right out of the hive, taking care not to rub the attached bees against their neighbours on the adjoining frame in so doing. Nothing tends more to irritate bees than to be rubbed between two combs.

The novice must not mind stray bees walking over his hands, and although this may give rise to considerable trepidation he must do his utmost to overcome his natural fears. Bees do not invariably sting whatever they alight upon, and they are just as likely to sting the frames, hive sides or anything else as the bee master's hands. One rule must be ever borne in mind. If stung don't flinch. As soon as possible, extract the sting. A capital way of doing this is to gouge it out with the nozzle of the smoker. Not only will the tarry matters which invariably collect round the nozzle of the smoker, due to the destructive distillation of the fuel, at once disguise the odour of the sting (this being a source of great irritation to the bees), but will also probably act as an antiseptic to the wound. The beginner must comfort himself that every sting he receives tends to lessen the effect of the next one, until in time he may reasonably hope to become sting proof.

But to return to our manipulations. The first frame having been withdrawn is lifted up level with the eyes and carefully examined. Accustom yourself from the first to note everything of interest. Look out for the queen. Observe the quantity of brood, whether sealed or unsealed, drone or worker, and the amount of sealed stores in the upper portions of the comb. The weight of the frame should be a guide to the amount of honey it contains. Also look out for queen cells, and if swarms are not desired break them off. The bees attached to

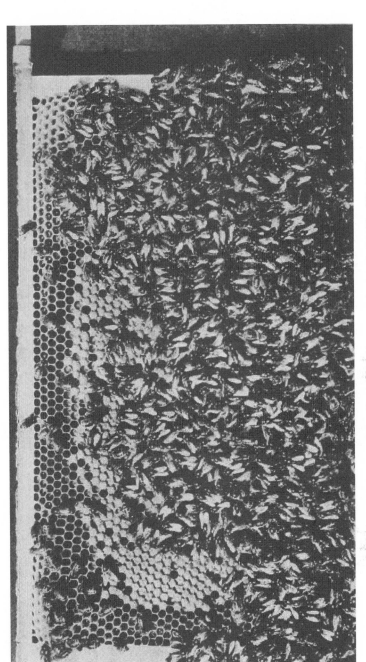

FRAME OF COMB FROM BROOD NEST, SHOWING SEALED BROOD AND BEES

(From a photograph by John C Douglas)

the frame can be removed by rapidly jerking the frame downwards from the eye level to within a few inches of the hive. Above all keep a sharp look out for indications of "foul brood" (see "Diseases").

Having examined one side of the frame, proceed to examine the reverse. To do this, take care never to hold the frame flat, as the comb is liable to break away owing to its great weight, but first lower the right hand corner until the top bar of the frame is vertical instead of horizontal. The left hand will now be immediately over the right hand and the frame will be pointing towards the left. Now revolve the frame for half a revolution, thus bringing the reverse side to view, and finally lower the left hand which will once again bring the frame into a horizontal position but upside down. These motions executed in the reverse order bring the frame back again into its normal position. Proceed in like manner until all the frames have been examined, giving a puff or two with the smoker from time to time as the bees become obtrusive.

After the examination is complete, take care that the frames are properly spaced and the dummy placed close to the last frame before closing up with the quilts.

If properly carried out, the previously described operations should have been performed without crushing a single bee. Like the sting odour, that emanating from a crushed bee tends to make the inhabitants of a hive irritable.

The beginner will derive considerable benefit if, before handling bees, he will carefully rehearse all the movements just described, using a hive fitted with frames, foundation, etc., in which no bees are housed. Particularly should he practise lifting the frames in and out without touching the sides of the hive or the adjoining frames, and also the movements requisite for examining both sides of the frames. This advice may seem at first sight trivial, but that it is not so the writer has amply proved.

Everything about the hive should be done quietly, without jerkiness or jarring, and all movements with the hands should be deliberate. A sudden, rapid movement of the hands over the exposed frames is almost certain to invite a sting. The hissing note of an angry bee needs only to be heard once to be readily recognised when heard again, the sound being totally unlike the peaceful hum of a bee in its normal condition. If an odd bee or two threaten, don't beat a hasty retreat, but stand firmly and put on as bold a front as is possible under the circumstances, when it will be found as often as not that, after one or two fierce darts, the enemy will retire.

Breath and perspiration are both obnoxious to bees. The remedy is obvious. This fact has long been recognised, and the advice so quaintly tendered by Butler more than two hundred and fifty years ago is so good that it is impossible to refrain from quoting it.

" If thou wilt have the favour of thy Bees, that they sting thee not, thou must avoid some things which offend them: thou must not be unchaste and uncleanly; for impurity and sluttiness (themselves being most chaste and neat) they utterly abhor: thou must not come among them smelling of sweat, or having a stinking breath, caused either through eating of leeks, onions, garlick, and the like, or by any other means, the noisomeness whereof is corrected with a cup of beer; thou must not come puffing and blowing unto them, neither hastily stir among them, nor resolutely defend thyself when they seem to threaten thee; but softly moving thy hand before thy face, gently put them by; and lastly, thou must be no stranger unto them. In a

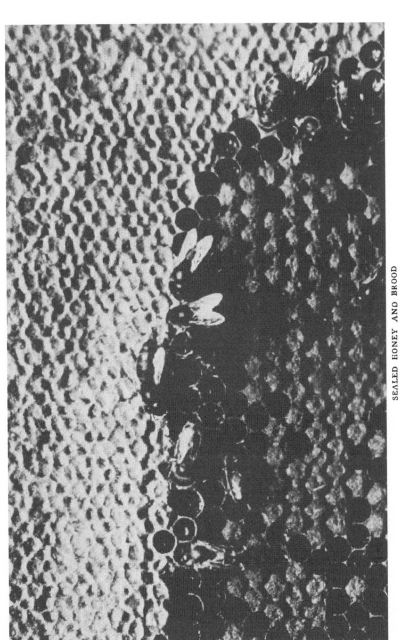

SEALED HONEY AND BROOD

(From a photograph by John C. Douglas)

word, thou must be chaste, cleanly, sweet, sober, quiet and familiar; so will they love thee, and know thee from all other."

Carbolic acid in lieu of the smoker has often been advocated as a quietener for bees. It is generally used by saturating a quilt, kept specially for the purpose, with a strong solution of carbolic acid. The method of preparing and the use of the carbolic cloth will be found fully described in sec. iii. chap. xvii., *Preparations for Winter.*

In removing stings, care should be taken not to press the poison sac; this only makes the sting more painful by injecting a further portion of the poison. Stings can be readily *pressed* out with the back of a knife blade, or gouged out with the thumb-nail. Most old beekeepers rub a little honey on the wound. This at least disguises the sting odour. Bathing in cold water, or a cold water bandage, will sometimes alleviate the pain. As a general rule, individuals in normal health suffer but little from the effects of an occasional sting, and it will usually be found that, once a genuine enthusiasm is aroused, "The labour we delight in physics pain."

CHAPTER VIII

As the season advances it will be noticed that the bees rapidly increase in numbers and activity, and a sharp look-out should be kept, by an occasional peep beneath the quilts, so as to prevent over-crowding and consequent preparations for swarming. To do this lift up one by one the four corners of the quilts for a second, so as to obtain a glance at the two outside frames. Should these be thickly covered with bees, drive them back with the smoker and add another frame to the most crowded side of the brood nest, and continue so to do until the full complement of ten frames is in the hive. To give room about a week in advance of the bees' requirements is one great step towards the prevention of swarming.

The brood nest being filled with combs, together with a large and ever-increasing population of bees, it soon becomes imperative to provide them with other means of accommodation. This is done by giving additional room over the brood nest, technically known as " supering."

The beekeeper should, by judicious stimulative feeding, so calculate matters that this crowded state of the brood nest should coincide with the honey flow. This means that he must make himself acquainted with the principal bee-forage of the district in which he dwells. A moment's consideration will show that to get a colony strong enough to take advantage of clover

—May or June—is a very different matter to getting one into condition for the heather harvest in August.[1]

Doolittle, an American bee-master of world-wide celebrity, lays great emphasis upon the importance of having a " full force of working bees in the field *at the right time.*"

To judge when honey is coming in, note the appearance of the upper portions of the combs in the brood nest. When these begin to whiten, owing to the bees elongating the cells, it may be assumed that the honey flow has commenced.

Another indication, at once apparent to the experienced eye, is the behaviour of the bees themselves. When honey begins to be obtainable in quantity, a notable increase in energy is at once observable.

Surplus honey is usually obtained in two forms— either in the form of comb honey or extracted honey. The beginner in beekeeping is advised to work for comb honey in the first instance, as involving less capital outlay ; afterwards he will take into consideration what particular form of honey is most in demand in the market for which he caters. The yield, per colony, of comb honey is less than that of extracted, the ratio being about 30 lbs. of the former to 50 lbs. of the latter. Comb honey, however, commands a higher price in the market.

[1] When heather alone is relied upon as the source of surplus honey, great judgment should be exercised in regard to stimulative feeding, as being likely to bring the bees too early into condition, so that by the time that the heather is in full bloom the queen will be diminishing her egg laying, and the population will consequently be decreasing.

CHAPTER IX

To produce section honey we require :—

1. Sections.
2. Foundation.
3. Section rack.
4. Dividers.
5. Queen excluder.

1. *Sections.*—These are now always made in one piece of American bass wood, and are 17 inches long by two inches wide, divided into four equal parts of $4\frac{1}{4}$ inches each, by means of equi-distant V-shaped grooves, these latter almost severing the wood, and each extremity is furnished with dovetails. These extremities when united, form a small box $4\frac{1}{4}$ inches square, with one dovetailed and three mitred joints. The full width of 2 inches is preserved at the corners only, the wood being cut away or incepted for $\frac{1}{8}$ inch top and bottom, or more generally in top, bottom, and sides, so that when a number of folded sections are packed together in the rack, passage-ways for the bees are provided in all directions. Sections incepted top and bottom only are termed two-bee-way sections, whilst those cut away on all four sides are called four-bee-way sections.

2. *Fixing Foundation.*—Sections are now obtainable, the top bar of which is split in two. In folding these, one half of the upper bar only is folded ; against this folded half place the foundation, and then press the

other half into position. The foundation is thus securely held in position, and the operation of fixing the same involves but little trouble.

Specially thin foundation known as " super " foundation should be used for the production of comb honey, and each section should be filled with a full sheet.

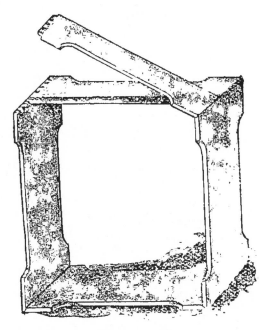

SPLIT TOP SECTION

The sections, after being folded and furnished with foundation, are placed (twenty-one of them, in seven rows of three each) in the rack. This is simply a box which encloses the sections on all sides, the bottom being fitted with girders or laths on which the corners of the sections rest. In order to ensure that the honey comb produced shall be of even thickness, it is necessary to use dividers or separators, thin strips of metal or wood furnished with suitable bee passages, betwixt each row of three sections. Finally, the whole (sections and separators) are firmly wedged into place by means of the " follower " and wedge.

SEPARATOR OR DIVIDER

To prevent the queen having access to the supers a sheet of excluder zinc should be placed over the frames in the brood nest. This has a series of longitudinal perforations of such a size that only the worker bees can penetrate. This should be placed over the frames in such a manner that the slots run at right angles to the top bars. It is recommended to place the zinc in absolute contact with the frames and not to have it mounted in a wooden frame. The excluder known as the " B.B.J." pattern is the best.

QUEEN EXCLUDING ZINC (FULL SIZE)

Premising that the honey flow has commenced and that supers are to be given, select a fine day when bees are flying freely, blow a little smoke into the hive entrance; next peel off the quilts, driving down the bees meanwhile with the smoker, place the excluder in position and then a rack of sections on the top.

It is most important that all be made warm and comfortable, so that no draught can possibly circulate in the super; strips of wood covered with cloth are useful, one on either side of the super, laid on the zinc covering the lugs of the frames in the brood nest. Of course the

SHALLOW FRAME SUPER

SECTION IN VARIOUS STAGES
(*From photographs by John C. Douglas*)

top of the section-rack is warmly covered with quilts, and it will be found an advantage to have, for the final covering, a specially long quilt, the ends of which can be snugly packed down at either end of the section-rack.

If a partially filled section can be placed in the centre of the rack it is a great inducement to the bees to commence work at once. Sometimes bees manifest considerable shyness in entering sections, the usual cause being cold and draughty supers; the remedy is obvious.

When the sections are about two-thirds full of comb and honey, another rack should be given *beneath* the first one. To do this use smoke, having the second rack at

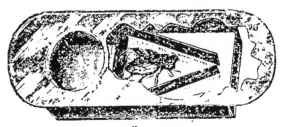

" PORTER " BEE ESCAPE

hand. "Prise" up the first rack at each corner in turn with a screwdriver, blow in a puff of smoke, then grasp the rack firmly with both hands; screw the rack round until it feels perfectly free, then, and only then, lift it clear of the hive, place it accurately on the top of the empty rack, replace both in the hive and pack up snugly as before. A third rack may be placed beneath the other two when it is judged that the second rack is being freely used. Probably by this time the first rack will be completed and ready for removal. The supers are of course crowded with bees, and the problem is to get rid of them in the most expeditious manner, and with as little disturbance as possible. The best way is to use a "clearer board" fitted with Porter bee escape, an ingenious device permitting the bees to pass through to the brood nest below, but effectually preventing their return upwards. Mr Meadows has devised a further modification, consisting of a tin slide whereby the bees may be readmitted to the supers at will. This device is especially useful when working for extracted honey. When using the clearer, first gently prise with a screw-driver the four corners of the super to be removed and blow in a little smoke, remove the super in the way

already described, place it on the clearer and immediately replace the whole on the hive. If this be done during the morning, the super will be found free from bees by the evening. If during the afternoon, the supers may be removed the following morning.

SUPER CLEARER

The sections will be found to be more or less firmly "propolised" together and difficult of removal from the rack. To remove them readily, procure three blocks of wood about 2 inches thick and of such a width and length as readily to fit in between the girders forming the supports for the sections. Place the rack of sections on these three blocks arranged in such a manner that they support the sections only. Then press down the rack, when the sections will be forced above the sides of the rack and can then readily be removed.

The completed sections now require to be cleaned from adhering propolis, those not completely filled or sealed over being returned to the bees to finish. Incomplete sections left over at the end of the honey season should have the honey extracted, the wet combs being returned to the bees to clean (in the manner to be described under the head of Extracted Honey), and the clean combs stored for use the following season.

CHAPTER X

EXTRACTED HONEY

THE actual method of working for extracted honey, as regards the putting on and the removal of supers, is in all respects the same as for section honey, the only difference consisting in the appliances used.

These comprise—

1. Shallow frame supers.
2. Shallow frames fitted with foundation.
3. Uncapping knife.
4. Extractor.
5. Ripener.

The supers are of exactly the same length and breadth as the brood-nest, but only 6 inches in depth, the frames being $5\frac{1}{2}$ inches deep, but in all other respects of standard dimensions. These frames require each to be fitted with ordinary brood foundation and W. B. C. ends, but wiring may be dispensed with if care is exercised when first extracting. Afterwards the combs become tougher, and will easily withstand the strain of extracting.

At the commencement of the honey flow, proceed in exactly the same manner as advised for sections, taking care to use the excluder zinc between the first super and the brood nest. When the bees are well at work in the first super, add a second one beneath the first one, and so on whilst the honey flow lasts. Do not remove any super until the combs are at least two-thirds sealed over. The longer the honey remains in the hive the more perfect the ripening.

Unripe honey, besides being thin, and lacking in flavour and aroma, is liable to ferment. Unsealed honey left over at the end of the season should be kept to itself after extracting, and may be used as syrup for feeding. Assuming the first super to be full of sealed honey, free it from bees by means of the super clearer and remove indoors for extracting.

The Extractor is a machine in which the combs (after removing the cappings) are placed, and the honey removed by centrifugal force, leaving the combs uninjured, and ready to be again refilled with honey by the bees. Considering the quantity of honey consumed by the bees when secreting wax, and the time taken in comb-building, together with the shortness of the honey season, it will be readily perceived that in the extractor we possess perhaps the most important appliance of modern beekeeping.

THE "COWAN" RAPID REVERSIBLE EXTRACTOR

As to the choice of an extractor, the reader is advised to consult the catalogue of some reliable bee appliance manufacturer. Both the "Cowan" and Meadow's Raynor extractor are perfectly reliable instruments, the former being fitted with an arrangement whereby the combs are automatically reversed without removing from the extractor. Multiplying gear is essential.

Before the combs are placed in the extractor, it is necessary to remove the cappings. Various knives have been devised for the purpose. An ordinary Christy bread-knife answers admirably. Whatever form of knife be used, it must first be immersed in hot water and wiped dry. To uncap, firmly grasp a frame by one end of the top bar, resting the other end on a dish, taking care that the comb is inclined outwards. Now commence with the warm knife to cut upwards, and the sheet of cappings will fall clear of the comb. Reverse the comb, and in the same manner uncap the opposite side. When two combs are uncapped, place them in the extractor. Seeing that the cells incline towards the top bar, the frames should be so placed that the bottom bar travels foremost. Experience will soon indicate the amount of speed necessary to thoroughly extract the honey. To avoid all risk of breaking new combs, only partially extract from one side at as low a speed as possible. Reverse and extract as much as possible from the other side, again reverse and complete the extraction from the side first operated upon. By this mode of procedure we avoid the pressure exerted by the full inner side of the comb upon the outer empty side.

After extraction, replace the combs in the surplus chamber, and when this is full, replace in the hive at night, when the bees are quiet. Withdraw the tin slide in the super clearer. The bees will ascend, and during the night thoroughly clean up the dripping combs. In the morning put in the slide, and by evening the super will be found clear of bees and the combs perfectly clean and dry.

Heather honey, owing to its greater consistency, is difficult to extract. In consequence, in heather districts it is better to secure the crop in the form of comb honey. If, however, extracted honey is required, the

combs are cut out of the shallow frames without previously uncapping, and the honey extracted by pressure,
the " Garstang " or the " Rymer " press being used for
the purpose.

The honey in the extractor will be found to contain
small particles of wax and to lack brightness in appearance. In order to make it presentable it requires

THE " RYMER " HONEY PRESS

straining. For this purpose a tall cylindrical tin-plate
vessel is used, fitted with a funnel and perforated gauze,
called a ripener. On the bottom rim of the funnel,
beneath the gauze, tie a piece of flannel freshly wrung
out of hot water. Place the funnel in position on the
ripener, and pour into it the honey from the extractor.
After all the honey has run through, the wax cappings
may be transferred to the funnel, so as to drain away
adhering honey. The honey in the ripener should be

kept at a temperature of from 70° to 80° F. for a few days. This assists the ripening process, and a layer

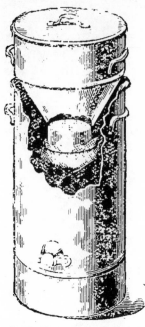

of thin unripe honey will be found floating on the top. This latter can be returned to the bees as food.

Extracting is best done in the evening, when the bees are all indoors, so as to avoid all chance of robbing and the consequent uproar. Similarly all vessels containing honey should be kept tightly closed, and great care should be taken to ensure that no particles of honey are spilt about.

Extractors are so constructed as to take to pieces readily, and, when finished with, every part should be carefully cleaned by pouring over it boiling water, and

HONEY STRAINER AND RIPENER

afterwards thoroughly dried. Avoid the use of galvanised iron in the construction of any receptacle for honey. Tinplate vessels only should be used, if metal be used at all. All honey valves should be thoroughly tinned.

DOUBLING OR STORIFYING

This is really working two stocks with one queen, and is carried out on the following lines.

Two strong stocks are selected, and from one of them all frames containing brood are transferred to a second brood chamber placed over the brood chamber of the hive to be doubled, super fashion. No bees are to be transferred but brood only. The vacant space in the deprived hive is filled with frames containing foundations, or empty combs, and the old stock is now in the condition of a newly-hived swarm.

The doubled hive will now rapidly increase in population by the hatching out of the two lots of brood, and the enormous number of bees engendered will quickly fill with honey the combs of the upper chamber as soon as vacated by the hatching bees.

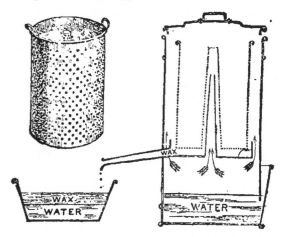

WAX EXTRACTOR

Experience will teach how much more super room will be required from time to time.

It is obvious that by this system the population of the hive doubled is speedily brought to its maximum, also since one queen, however prolific, cannot do the work of two, the population just as quickly subsides. Therefore the process of doubling should be so carried out that the hive is at its greatest strength simultaneously with the main honey flow.

Wax Extracting

If much extracting be done, instead of using the strainer of the ripener as a receptacle for the cappings, it is perhaps better to provide a vessel specially for the purpose. An ordinary potato strainer answers the purpose admirably.

At the end of the season all the cappings, together with old combs and all accumulated fragments of wax, are melted down. This is done in a Gerster's wax extractor. The wax to be extracted is placed in a perforated metal cage, which fits into an outer case, and the whole fits tightly over a pan containing water. This is placed on the fire. When boiling, steam circulates around the wax in the upper part of the apparatus, the molten wax being caught in a tray immediately underneath the perforated cage, whence it is conveyed to a dish (which should contain a little hot water) by means of a side delivery tube.

CHAPTER XI

THE prosperity of any given colony of bees is entirely dependent upon the queen or mother-bee, and should she be lost through any cause the colony gradually loses heart, dwindles away, and finally becomes extinct.

Queenlessness may result from a number of causes, untimely manipulation in the early spring or late autumn being probably the chief cause, any untoward disturbance at such times causing the bees to cluster closely around the queen in a solid compact mass, known to bee-keepers as " balling." A queen may be lost during her mating trip, through inability to find her way home, or, as very seldom happens, she may be caught by some bird.

Queens are sometimes lost during swarming from sheer inability to fly. The loss of a queen is soon detected by an observant beekeeper, the bees of the queenless stock behaving in marked contrast to their more fortunate neighbours. The bees are seen to run in and out of the hive, and up and down the flight-board, as if seeking for something, those bees who actually take flight only flying for a short distance and speedily returning. Should other colonies be carrying in pollen freely, another indication of queenlessness is afforded by the fact of the queenless stock carrying in little or no pollen. A queenless stock will also tolerate drones at a time when other stocks have driven them out.

Absence of eggs and brood is also indicative of

queenlessness, but usually outside appearances form sufficiently conclusive evidence.

Should a stock be found to be queenless it should either at once be supplied with a new queen or else united to another stock, according to circumstances. (See "Uniting.")

CHAPTER XII

SHOULD a colony become queenless at a time when neither eggs nor brood are present, it soon becomes extinct unless supplied with a new queen. If, however, the hive contain eggs or brood the bees set about replacing their lost mother by constructing queen cells, but before so doing valuable time is lost in fruitless search for their missing queen. Hence it is more than probable that by the time the bees fully realise their loss, all larvæ in the hive will have advanced considerably in development, and queen rearing operations will perforce have to be commenced by specially feeding larvæ which may by this time be fully three days old.

Experience has shown that queens thus raised are not so prolific or so vigorous as those raised from eggs which have been destined from the start to produce queens. Therefore in raising young queens the bee-keeper should always take care that they are raised from eggs and not from larvæ. The system of queen rearing devised by Mr H. W. Brice fully accomplishes this object, and is best described in his own words.

" Choose the best stock in the apiary, remove the queen and sufficient bees and brood to form a good nucleus, giving them one empty frame of comb, and allow the original stock to raise queen cells. This they will do first in their way and afterwards (if my advice is followed) in the way the beekeeper desires.

" (1) Shortly after the queen is removed the bees will

be greatly distressed; (2) as soon as they realise that she has gone, many eggs will be destroyed or removed, and in about three days cells will be started in hot haste over larvæ now too old to produce good queens. (3) Now comes the time to do it yourself! Go to the original stock and cut out all queen cells, and from the nucleus remove the once empty frame of comb you gave them now filled with eggs just hatching. Give this to the stock right in the centre of the brood, and you will see cells arise on larvæ almost as soon as hatched from the eggs. These will eventually produce queens far superior to those the bees would have raised in the first instance. Everything is in their favour, age of grubs, number of nurse bees, and above all the earnest desire of the bees themselves to produce queens. On the fifth day after the frame of eggs has been given examine the hive and remove all cells from any other frames save the selected one. Ten to twelve days after the giving of such frames, divide the stock up into nuclei, giving a cell to each divided part. Finally when the new queens are hatched and mated and laying (generally within fourteen days of their hatching) replace all old queens with the new, rejoin all the nuclei into one stock, and the apiary is requeened with queens worthy of the name."

A nucleus is a small colony of bees, usually of three combs, which may be placed in an ordinary hive, contracting the brood nest to the required size by means of the division boards; or small hives specially built may be used.

To form a nucleus, select a prosperous colony and from it remove three frames, taking great care that the queen is left in the parent hive (excepting of course in the first stage of queen rearing just described). One of the frames should contain unsealed brood and eggs and the other two honey. Place these together with

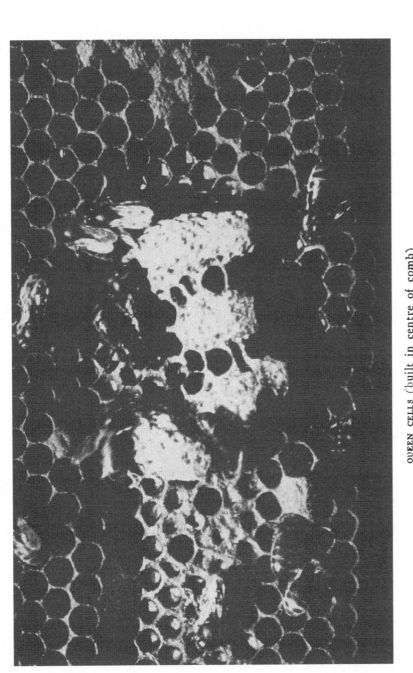

QUEEN CELLS (built in centre of comb)

(From a photograph by John C. Douglas)

all adhering bees, in the nucleus hive and shake in more bees. These operations should be carried out at a time when the bees are flying freely, so as to secure in the nucleus as many young bees as possible; these never having previously been abroad, will in their first flight mark the position of the nucleus and return, whereas all the old bees included in the nucleus will fly back again to the parent hive. A separate nucleus will be required for each queen cell. Queen cells require very careful handling. Frames containing such should never be jerked. All adhering bees should be brushed off.

The queen cell should be cut out from the comb with a sharp thin-bladed pen knife, taking care to cut away with it a piece of the comb to which it is attached at least the size of the queen cell itself. Throngh this insert a stout needle or pointed wire about $1\frac{1}{2}$ inches long, and place the cell betwixt two top bars of the nucleus. The wire will form a bridge from bar to bar and the cell will be suspended between the combs in the warmest part of the hive. Cover all snugly with the quilts, and in a short time the bees will have still further secured the cell.

The following data will be found useful—

The egg hatches on the *fourth* day.
The queen cell is sealed on the *ninth* day.
The queen emerges on the *sixteenth* day.

Therefore the young queens may be expected in less than a week's time after inserting the cells, and in about another week, under favourable conditions, the young queens should be successfully mated. In order to ascertain if such be the case, examine the combs and if eggs are present all is well. The queens may be kept in the nuclei until required elsewhere.

Select only the very best stocks for queen-rearing

purposes, the queens of such should not be more than two years old. It may be taken as a general rule that disposition is transmitted by the drone; working qualities and general vigour by the queen. In cross-breeding, in order to secure bees of gentle disposition (according to Simmins), the foreign element should come from the male, the virgin queen being pure native. This fact may possibly explain the many conflicting opinions expressed regarding the disposition of hybrid bees.

In order to secure, as far as possible, that the young queens are fertilised by drones from the particular stock required, it is necessary to ensure the presence of drones in such stock, in considerable numbers at a time when drones are either but few in number or entirely absent from other colonies. Therefore it behoves the bee-keeper to make preparations for queen rearing early in the season. Stimulative feeding of the stock required to produce the male element should be commenced in April, frames containing brood may be added from other colonies, and when the hive begins to be crowded insert in the centre of the brood nest a frame of drone comb. By this means we secure an abnormally large number of drones early in the season.

When the drones begin to hatch commence to rear queen cells in the manner before described.

CHAPTER XIII

INTRODUCTION OF QUEENS

HAVING successfully raised the requisite number of queens, the next stage is to introduce them to the colonies, the queens of which we are desirous of superseding. This requires care, otherwise the bees will destroy the stranger. If the alien queen be enclosed in a cage, and the whole introduced into the hive, the bees gradually become reconciled to the presence of the stranger, and if the new queen be released in forty-eight hours she is generally accepted.

But perhaps the simplest as well as the safest method of introducing stranger queens is that known as Simmins' direct queen introduction.

Before re-queening, the first step obviously is to find the old queen we desire to supersede. To do this, open the hive at a time when the bees are flying freely, lift out the central comb and carefully examine both sides of the comb, paying particular attention to any particularly dense cluster of bees. Queens are extremely shy, and speedily bury themselves beneath a cluster of the workers. Should she not be found in the first frame examined, hang the frame on a comb stand, or place it in a spare hive, and search each frame in turn until she is found, when, seizing her by the wings, either kill her at once or otherwise dispose of her. Should a thorough search fail to discover the queen, queenlessness may be suspected. To set all doubt at rest, insert a comb containing brood, if the bees commence queen cells thereon, no queen is present. Failing this, the queen must again be sought for on the first favourable opportunity.

Direct introduction is carried out as follows :—

The new queen is to be inserted in the evening of the day on which the old queen has been removed, and previous to introduction she is to be kept *quite alone* and *without food* for at least thirty minutes, taking care that she be not chilled meanwhile. A short test tube is handy for the purpose, and should, for the half hour it is occupied by the queen, be kept either within the vest or in the trousers pocket.

After darkness has set in, proceed to the queenless hive—with a lamp if necessary—lift up the quilt from one corner, drive the bees back with a little smoke, and let the new queen run in. Cover all up at once, and on no account interfere with the hive for fully forty-eight hours.

Fertile workers.—It sometimes happens that in the case of a queenless colony one or more of the workers commences to lay eggs. Huber supposes that these fertile workers are bees who have been reared in the cells nearest to queen cells, and whose ovaries have been somewhat more fully developed from their partaking of " royal food " accidentally placed in these cells. Whatever may be their origin they are a great pest. Of course a fertile worker is only capable of laying drone eggs. These pests are easy of detection, the eggs being laid in the most irregular manner here and there over the surface of the combs.

The best way to deal with such colonies is to unite them to one or more other colonies having a fertile queen.

CHAPTER XIV

DRIVING has for its main object the removal of bees from straw skeps in order to obtain the honey therein without sacrificing the lives of the busy workers. At first sight, this would seem to be an operation somewhat outside the scope of modern beekeeping, but cottagers who keep their bees in old-fashioned hives, and who yearly "drown" the inhabitants of those hives they wish to take, are still to be found in most districts. Many such are only too glad to give the condemned bees to anyone who cares to take them.

The beekeeper, during his first season, should seize upon any opportunity that arises for driving. The operation is interesting, and if carried out carefully with all due precautions, perfectly safe ; and a successful drive is calculated to inspire one with increased confidence in handling and manipulating bees. The bees so obtained may be united to any existing colony that is at all weak, or if spare *worked out* combs are in hand containing sealed stores, a new colony may be made up from two or three lots of driven bees united in one hive, provided always that the bees are free from disease.

The apparatus required comprises :—

Driving irons. These may be made out of two pieces of iron or steel wire, about $\frac{1}{8}$ inch in thickness, bent in the manner shown in the figure, the distances between the points should be 12 inches, and the points themselves about 1 inch in length.

A few sharp skewers of the same strength as the irons, and about 6 inches in length. Have these made with good large loops so as to be able to grip them well. It is well to be provided with half-a-dozen of these as they are exceedingly liable to lose themselves at critical moments.

The remainder of the outfit should comprise the smoker, together with plenty of fuel, veil, gauntlets, a bottle of thin, warm syrup, one or two empty skeps to contain the driven bees, with squares of cheese cloth large enough to tie the four corners over the top of the skep. These latter, if folded and wrapped round the syrup bottle, will keep it warm for hours. Always

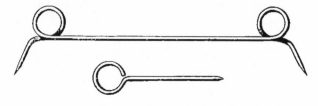

DRIVING IRONS AND SKEWER

make a point of taking with you everything you are at all likely to require, and never rely on getting at your destination, even brown paper for the smoker. A fine, warm afternoon, when the bees are freely flying, is generally supposed to be the best time for driving.

Should the hives composing the cottager's apiary be very close together, and should the bees appear at all excitable, it is a good plan, before commencing operations, to close the entrance of all the hives with grass pushed in *lengthways* (excepting, of course, the hive selected for driving) so as to imprison the bees without depriving them of air. Next blow a little smoke into the hive about to be operated upon, and in a few seconds give a few slaps with the flat hand around the hive. This so startles the bees that they manifest no disposition to sting. After a second puff or two of smoke and a few minutes' wait, turn the hive bottom upwards. The bees will be found clustered in the crown of the hive and scarcely a single bee will fly out. To make security doubly secure, a little of the warm syrup may be poured over the combs, and the hive replaced upon its stand for a minute or two longer, to give the bees time to lick up the syrup. Then again overturn the hive and at once carry some little distance away from its fellows; place it on the ground or anywhere handy, still bottom upwards, and with a couple of the skewers fasten one of the spare skeps in such a manner as to form a dome over the upturned hive. The skewers should be inserted from either side of the hive, but only an inch or two apart, so as to form an extemporised hinge. Next take the driving irons, one in each hand, and insert the lower point of each into the sides of the bottom hive, then raise the upper hive on its hinge until it forms an angle of about forty-five degrees, and secure it in this position by squeezing in the upper points of the driving irons. These operations should be carried out as expeditiously as possible, and the novice should practise on two empty skeps previous to driving.

One most important point is that the combs should run from front to back, *i.e.* the combs should be at right angles to the two skewers forming the hinge, in the manner shown in the illustration.

With your back to the strongest light, commence to beat on the sides of the hive with both hands, smartly and firmly, in such a manner as to jar the combs, but not sufficiently strongly as to run any risk of breaking them away from their fastenings. The blows should be rhythmical, both hands striking the hive at the same time, and the intervals should be about the same as between the footfalls of an ordinary walker. Very quickly the bees will begin to roar, and will move in a procession from the lower to the upper empty hive,

so that in a very short time the lower hive will be entirely
empty of bees.   This operation may take from three to
twenty minutes.   Whilst the bees are running up, look
out for the queen, and if necessary seize her.   She may
be stored in an empty matchbox.   It is better, as giving
a direction to the bees, to slightly tilt up the lower hive,
the highest point being where the two hives are joined.

After removing the hive to be driven from its stand,
some operators place an empty skep on the stand to
catch all bees returning from the fields.   The writer
generally places a square of cheese-cloth or calico on the
old stand and, immediately the bees are driven, puts the
skep containing the driven bees on the cloth.   In a few
minutes the majority of the flying bees rejoin their
companions, then the cloth is tied by its four corners
over the top of the skep and afterwards still further
secured by tying string around the rim.

One or two precautions are necessary :—

1. As driving always takes place after the close of
   the honey season, be particularly careful not
   to spill syrup about, otherwise robbing and
   fighting will speedily follow.

   The deprived hive containing its combs and
   store of honey should be removed indoors as
   speedily as possible for similar reasons.

2. As the point of contact between the two skeps
   is comparatively small, the rush of bees is very
   likely to be much greater than this narrow
   bridge can well accommodate, and in consequence a considerable number of the bees on
   either side of the junction run down on the
   outside of the skep.   In order to avoid this,
   tie a handkerchief from iron to iron, stretching
   it moderately tightly, or better still, have the
   two irons connected with a piece of calico 12
   inches wide and about 30 inches long.   Imme-

SKEPS ARRANGED FOR DRIVING

(From a photograph by John C. Donglas)

diately before inserting the irons into the skeps, turn one of them upside down in such a manner as to make a fold in the centre of the calico. This will effectually prevent the escape of any bees in passing from the lower to the upper skep.

3. In cold weather, when driving never neglect the use of warm syrup, and give the bees ample time to lick it all up. By this means the temperature of the hive is raised considerably higher, and the bees become in first-rate condition for driving.

CHAPTER XV

UNITING—ROBBING

Any stocks found weak in the autumn should be united to form one or more strong stocks. Such united stocks winter well, and consume, *pro rata*, less food than weak ones. Hives to be united should be gradually moved close together, moving one or the other, as the case may be, only on fine days, and then not more than three feet at a time. When close together, remove the queen from one of the hives, together with all frames destitute of brood. Space the remaining frames as widely apart as possible, and liberally dust the bees with flour from a common flour dredger. Between each pair of widely-spaced frames, place a frame from the other hive, having first dusted the bees with flour in like manner. Remove the empty hive, and the operation is complete.

Bees deprived of comb, brood, and stores will readily unite with others in a similar plight, the queens fighting for supremacy.

Instead of alternating the combs from the hives to be united, the bees can be shaken off on to a board (arranged as for hiving swarms) placed in front of a third hive, when, as a rule, they will run in peaceably. Should any signs of fighting be seen, smoke the bees vigorously.

Swarms readily unite, as also do driven bees.

ROBBING

In the springtime, before honey begins to come in, and in the autumn, after the honey flow has ceased, bees show a very decided tendency to rob their weaker neighbours. Robbing, when once thoroughly started, is most difficult to stop; therefore prevention is better than cure. A bad case of robbing once experienced is not likely to be soon forgotten. The whole apiary is speedily in an uproar, and with their lapse of honesty the bees seem to lose all other good traits, flying about in all directions and stinging everybody and everything within reach.

Strong colonies will usually be able to hold their own, but it is well to contract all hive entrances somewhat in the spring and autumn.

Take great care not to drop any wax, honey, or syrup on or about the hives. When replenishing feeders, always have a wet sponge handy, and instantly wipe up any spilt syrup.

Do not at such seasons open hives unless absolutely compelled, or extract honey during the day time.

When robbing has once set in, contract the entrance of the attacked hive to one bee space, and place on the alighting board, at either side, cloths saturated with crude carbolic. A sheet of glass reared in front of the entrance often effectually baffles intruders. In extreme cases, the hives may be completely covered by throwing over all a sheet saturated with weak carbolic solution.

Bees never rob when forage is plentiful.

CHAPTER XVI

THE MARKET

THE greatest possible care should be taken that all honey intended for the market is put up in as attractive a form as possible. This is a point of the utmost importance, the neglect of which has done

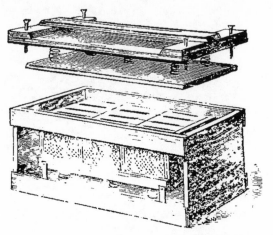

THE "COWAN" TRAVELLING CRATE FOR SECTIONS

much to prejudice grocers and others against stocking honey. Weeping sections, and sticky, messy bottles are an abomination, no matter how excellent the honey *per se* may be.

All sections should be glazed after cleaning, and graded according to weight and finish, the short weight sections being either disposed of locally at a reduced rate or used for home consumption. Those only partially filled should be returned to the bees to be cleaned

up, and will then form a most valuable asset in the
following spring in the shape of partially worked out
combs to be used as baits to induce
early work in sectional supers.

The before - mentioned glazing
consists in enclosing the section
between two plates of glass, the
glass being kept in position by
means of a strip of lace - edged
paper specially made for the purpose and sold by the appliance
dealers. The sections thus become for all practical purposes
hermetically sealed.

Sections are very fragile, but if
required to travel any distance by
rail they are best sent packed in the
"Cowan Travelling Crate." These
crates are very cheap, and are made to contain either

HONEY BOTTLE

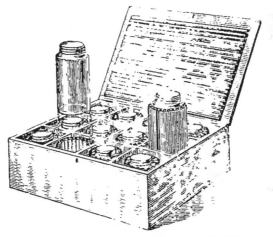

THE " COWAN " TRAVELLING CRATE FOR BOTTLES

twelve or twenty-four sections. The sections being
supported by sofa springs on all sides travel securely.

Extracted honey, if not disposed of in bulk, is usually put up in 1 lb. metal screw-capped glass bottles. Each metal cap contains a wad of cork, and if this be immersed in molten paraffin-wax or bees-wax previous to use, but little fear need be entertained of honey leaking. The whole should be neatly labelled, the label showing clearly the name and address of the producer.

For sending away honey in bulk, most appliance dealers supply special tins, either round or square, holding from 2 lbs. upwards to 56 lbs. The square tins, although more expensive, pack better and take up less room than the round ones.

CHAPTER XVII

PREPARATIONS FOR WINTER

AFTER the honey flow has ceased, and all supers have been removed, the queen excluder may be dispensed with. As it is generally found firmly propolised to the top bars of the frames, its removal without infuriating the bees seems almost impossible. But if the excluder is used without being mounted in a frame, its removal is not difficult. Procure a piece of calico 18 inches square, and moisten it with a solution of one part Calvert's No. 5 carbolic in two parts of water, taking care to keep the acid from contact with the skin. This thrown over the frames is an effectual quietener of bees, and was used by its inventor, the late Rev. G. Raynor, in lieu of the smoker in all manipulations; but it is especially adapted to the removal of a firmly propolised excluder zinc. It not only mechanically prevents the bees boiling up, but the odour of the acid drives them downwards, and it leaves *both* hands at liberty. Armed, then, with the carbolic cloth and a long-bladed knife, proceed as follows: Remove all quilts, and replace with the carbolic cloth. Lift up one corner of the excluder and insert the knife, forcing it along the top of each frame, at the same time gently peeling away the zinc. Having removed the excluder, take the opportunity of cleaning the top bars as much as possible. The old propolised quilt should be burnt, and a clean one (with a feed-hole) put in its place.

Breeding should be kept up by gentle, stimulative

feeding, taking every precaution to avoid starting robbing. About the second week in September, go over all the stocks, and any containing less than from 25 to 30 lbs. of sealed stores should have their stores augmented by means of the rapid feeder without delay. It is important that all winter food be sealed (see Dysentery). Any not sealed should be extracted. All weak stocks should be united and fed up to right condition for wintering.

As an additional precaution against starvation, give a cake of candy on the top of the frames. Where candy is not given, place a couple of sticks about $\frac{3}{4}$-inch square across the frames at right angles to the frames, so as to provide a passage way from comb to comb. When candy is used, the bees eat out their own passages.

Over all put two or three calico quilts, a similar number of flannel ones, and lastly a chaff cushion. This latter is a bottomless box 4 inches deep, and the same size as the brood-chamber. Tack on a bottom of calico, and fill up with dry chaff or cork dust. See that all roofs are perfectly water-tight.

When there is no longer any danger of robbing, open the entrances to at least 6 inches wide. Remove accumulated dead bees from time to time with a hooked wire, and when snow is on the ground, rear a board in front of each hive, so as effectually to shade the entrances. Otherwise the glare will entice the bees to fly, only to perish of cold.

All hives and appliances not in use should be thoroughly cleansed and scrubbed out with carbolic soap at once, after which they may be stored for use the following season.

SECTION IV

CHAPTER I

DISEASES

FOUL brood and dysentery are the only two diseases at all likely to cause serious trouble in the apiary; the former especially so, and it behoves the beekeeper to exercise the utmost vigilance to prevent his stocks from becoming contaminated. Should such unfortunately occur, it is important to detect the disease in its earliest stages, as once the disease gains a firm footing, to eradicate it is well nigh impossible.

Foul brood is due to the presence of a distinct micro-organism named *bacillus alvei*, which in a diseased colony, is found not only in the brood, but also in the bees and queen. Under favourable conditions as regards nutriment and temperature, two generations can be raised in an hour, or one single parent can give rise to no fewer than 336 generations in one week, each individual bacillus thus produced again multiplying itself at a similar rate.

The bacilli having increased to such an extent as to exhaust all the nutriment upon which their existence depends, change in both form and character, turning into spores, in which state they lie dormant, preserving their vitality for an indefinite period, successfully resisting changes of temperature, from extreme cold up to a heat of 300° F. Chemical reagents which would readily destroy the bacillus are without effect upon the spores, unless administered in such concentrated condition as to destroy the bees.

Here is where the real danger of foul brood comes in. Once the bacilli sporiolate, successful cure of a diseased colony becomes exceedingly difficult.

Needless to point out, foul brood is terribly infectious.

On examining a comb taken from a healthy hive, the brood will be seen to be of a pure pearly whiteness, and to lie curled up in the cells somewhat in the form of a letter C. Diseased larvæ, on the other hand, quickly lose their plump appearance, and gradually extend themselves in a horizontal position. The colour changes from the characteristic pearly whiteness to a pale yellow, which deepens more and more in colour as the disease advances, until finally the larvæ decompose, leaving a dry brown scale.

In the case of sealed brood affected, the cappings are generally darker in colour, more or less concave or indented, and sometimes perforated. On probing with a pin or match, these cells will be found to contain a sticky coffee coloured ropy mass of putrescence, which generally smells most offensively, the odour somewhat resembling that of putrid fish.

Any instrument used for probing should at once be destroyed. It is a good plan to have a bowl of carbolic acid at hand when examining stocks for foul brood. Care must be taken clearly to differentiate between foul brood and chilled brood. Larvæ affected with foul brood turn *yellow*, then *brown*, and finally leave a *brown scale*. Chilled brood turns *grey*, and afterwards nearly *black*.

To successfully treat an infected colony, two objects must be kept in view. Firstly, to get rid of all spores; and, secondly, to destroy the bacilli. Should the disease fortunately be detected in the very earliest stages, no spores will be present, and simple feeding

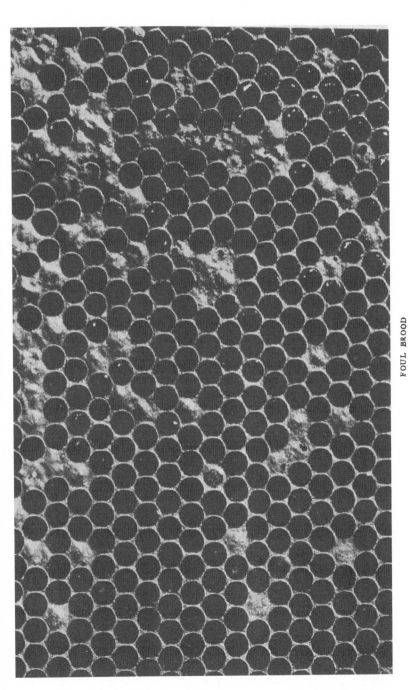

FOUL BROOD

(From a photograph by John C. Douglas)

with a properly medicated syrup will sooner or later effect a cure. The disease having advanced to the sporiolating stage more drastic treatment is called for.

The bees of the infected stock should be shaken into a straw skep and kept confined therein for about a couple of days. In the meantime, burn the infected combs, frames, and quilts, and with a painter's blow-lamp thoroughly scorch all internal parts of the hive, not forgetting the floor board, after which scrub out thoroughly with strong carbolic solution, and when dry give the hive two good coats of white-lead paint, inside and out, with the object of imprisoning any spores which may have escaped destruction either by burning or by the action of the carbolic solution.

The bees in the skep are now run into a clean hive fitted with full sheets of foundation, and a new queen from a reliable source introduced. Feed with syrup medicated with napthol beta (*see* Stimulative Feeding) to encourage brood rearing.

The skep in which the bees were confined should be at once burnt.

The burning of the frames, etc., gets rid at once of the majority of the spores. Any spores conveyed mechanically by the bees into their new home, will, in a very short time, find conditions suitable for their germination, only to be destroyed by the action of the napthol beta in the syrup.

Prevention is better than cure, therefore always feed in the spring and autumn, or whenever necessary, with *medicated* syrup. Also keep a little napthaline in the hives (one ball split in two being the proper quantity for each hive). Place each half ball in the corner opposite the entrance.

Points to be remembered in dealing with foul brood.—Foul brood is just as infectious as small-pox, scarlet fever, or cholera.

Never manipulate infected and healthy stocks indiscriminately.

After handling a diseased stock, wash with carbolic soap, and spray the clothing with weak carbolic.

Destroy all probes immediately after use.

Remember that although it is comparatively easy to destroy the bacillus, it is well nigh impossible to destroy the spores in the presence of live bees and brood.

Hopelessly infected stocks should be burnt, lock, stock, and barrel.

Honey taken from infected hives, although containing both bacilli and spores, is not spoilt for consumption, but great care should be taken that no bees gain access to it. Repeated boilings at intervals will ultimately sterilize it, but even, after being so treated, it is not wise to use such honey for bee-feeding.

Formalin is being used by the expert of the Irish Congested Districts Board for the treatment of stocks affected with foul brood.

A large hole is made in the floor boards, covered with perforated zinc on the upper side, and a sliding shutter beneath. A sponge, saturated with a 10 per cent. solution of formalin, is placed within the cavity.

Twenty-six stocks, affected with foul brood in the summer of 1900, were isolated and treated continuously with formalin. Examined in the early summer of 1901 foul brood was discovered in only five of the stocks. The healthy stocks, when again examined during the summer of 1902, were apparently still healthy.

DYSENTERY

Dysentery usually manifests itself during winter and early spring, and is generally caused by the bees feeding on unripe, unsealed, or fermented stores. Hence the advisability of feeding up all stocks in good time, so that they may ripen and seal all stores before the cold weather sets in. As a precautionary measure, it is a good plan to extract all unsealed stores prior to packing up for the winter.

Bees suffering from dysentery discharge their excrements over the combs, the inside of the hives, and around the entrance, the voided matter being dark brown in colour, semi-liquid, and of an offensive odour.

To cure, transfer the bees to a clean hive (placed on the original stand), give fresh combs of sealed stores previously warmed somewhat, and cover all up snugly with warm, dry quilts, when usually a speedy cure is effected.

Note.—Bees, after long confinement during inclement weather, take the earliest opportunity of indulging in a " cleansing flight," signs of which may be observed on hive roofs and surrounding objects, but very little around the entrances. This natural process must not be confounded with dysentery.

CHAPTER II

ENEMIES AND PESTS

THE WAX MOTH, the larva of which eats its way through the mid rib of the combs, leaving in its track long web-like tunnels, is the pest most to be feared. Strong colonies rarely suffer from an attack. Weak colonies, and especially empty stored combs, need a watchful eye.

Empty combs, stored from one year to another, should be so enclosed as to prevent the access of the parent moth, fumigating them with burning sulphur previous to storage.

Should the larva make its appearance, the above treatment with burning sulphur will eradicate it.

When examining hives, keep a sharp look-out amongst the quilts, along the top bars of frames, and along the saw cut, and destroy any grubs found lurking.

Ants are sometimes troublesome, especially when spring feeding, congregating in large numbers around the feeding stage. They do but little real injury. To prevent their gaining access, stand the legs of the hive in saucers of water or, better still, oil.

Mice occasionally gain access to hives and do a considerable amount of damage. Hive entrances should be sufficiently shallow to prevent their access. Hives having deep entrances should have a wire stretched across.

Wasps sometimes play havoc with a weak colony. Strong colonies may be relied upon to defend themselves. All wasps seen flying in the early spring are queens and should be destroyed, each wasp so destroyed being a potential colony.

Birds.—The blue-tit at times acquires the habit of feeding on bees, especially during the winter months. His methods are decidedly ingenious. He alights upon the flight board and taps with his beak, which procedure quickly brings one or more bees to the entrance, only to be immediately pounced upon by the expectant bird. Stern methods are to be deprecated. Old fish-net is cheap, and a piece hung in front of a hive during the winter will not materially interfere with the bees during occasional cleansing flights.

Toads should be kept at a respectful distance, and not allowed to lurk beneath hives.

Blind Louse (Braula Caeca) is a small reddish-brown wingless insect about one-sixteenth of an inch long, and sometimes found infesting bees, particularly the queen. They cause no serious damage, and their presence need not alarm the beekeeper.

H

CHAPTER III

BEES when taking their first flight carefully mark the situation of their hive, and having once taken their bearings, apparently never trouble themselves further, but return from whatever direction they may have flown with unerring instinct. Therefore, whenever it becomes necessary to move bees from one situation to another, special precautions must be taken to ensure the bees marking the change of locality.

If bees be moved a distance of two miles and over, the change of locality is at once apparent to them, and the new situation is carefully marked. Swarms invariably take fresh bearings, as also do bees when reduced to the conditions of a swarm, as, for instance, after driving.

Should it be necessary to move stocks a short distance in the spring, they should be moved one or two feet each day, *when the bees are flying freely,* until the desired site be reached.

When bees have been confined for a period of fourteen days or over by stress of weather (as in winter), they may be moved short distances with comparative safety.

Before moving a hive bodily, the combs in the brood nest must be secured in such a manner as to prevent their swinging about and crushing the bees, and care must be taken that complete ventilation is provided, as the excitement set up causes a considerable rise of temperature. Remove all the quilts but one, then screw flat strips of wood across the frames (over the

114

metal ends) to the brood nest.   These will prevent the
frames from rocking.

Straw should be packed in the air space between the
brood chamber and the brood lift, so as to keep the
latter in its position, and either close the entrance with
straw (which answers very well for an hour or two's
journey), after having previously removed the shutters,
or else close the entrance with perforated zinc.   Perforated zinc may also with advantage replace the one
quilt left over the frame, especially if the bees have to
travel any considerable distance.   The outer brood lift
may be secured to the floor board with a couple of
screws through the plinth.   A rope passed beneath the
floor board and over the roof makes all secure.

Old combs, being tough, will bear a journey much
better than combs of the present season.   With the
latter it is hardly safe to move a stock, unless it can be
carried by hand.

CHAPTER IV

MEAD

MEAD, meath, metheglin, hydromel, all of which terms may be considered as synonymous, is a beverage prepared by the fermentation of honey, the potency of which varies according to the amount of honey used in its production.

Previous to the introduction of malt liquors mead was the universal drink of northern nations, and was held in the highest estimation—even Ossian sings its praises.

Bevan in his " Honey Bee," however, holds it in but light esteem, and considers that our ancestors' liking for this beverage must have proceeded " either from their unpampered simplicity of taste, or from their having a better method of making their mead than has been handed down to posterity."

The truth of the matter probably lies in the fact that much of the mead met with at the present time (and probably more so at the time Bevan wrote) is made by cottagers from the worst of the honey which can be used in no other way, mixed with the rinsings from old, black, and often mouldy pollen-choked combs, whereas our fore-fathers would probably use the best of the honey at their disposal for the brewing of their favourite drink.

Cassell's " Dictionary of Cooking " gives the following methods of preparing mead :—

" Let the whites of six eggs be well incorporated with twelve gallons of water, to which add twenty pounds of honey. Boil these ingredients for an hour, then put into the liquor a little ginger, clove, cinnamon and mace, together with a small sprig of rosemary. As soon as the liquor is cool, add a spoonful of yeast, and pour the mead into a vessel, which should be filled up while it works. When the fermentation ceases, close the cask and deposit it for six or eight months in a vault or cellar of an equal temperature, in which the liquor will not be liable to be affected by the changes of the weather. At the end of that time it may be bottled, and will then be fit for use.

" A more simple, and to some palates more agreeable method, is to mix the honey in the proportion of one pound to a quart of water, which is to be boiled, scummed and fermented in the usual manner, without the addition of any aromatic substances. It ought to be preserved in a similar manner, and bottled at the expiration of the same time."

The Rev. Mr John Thorley, in his "Female Monarchy" (1744), describes—

"How to make Mead, not inferior to the best of foreign wines :—

" Put three pounds of the finest honey to one gallon of water, two lemon peels to each gallon ; boil it half-an-hour (well scummed), then put in while boiling lemon peel. Work it with yeast, then put it in your vessel with the peel, to stand five or six months, and bottle it off for your use.

" N.B.—If you choose to keep it several years, put four pounds to a gallon."

Sir J. More, writing in 1717 " Of the Husbandry of Bees, and the great Benefit thereby," distinguishes between two kinds of Mead :—

" Meath or hydromel is of two sorts, the weaker and the stronger meath, or metheglin. Mead being made from honey and water ' strong enough to bear an egg the breadth of a twopence above the top of the liquor.'

" Metheglin is the more generous and stronger sort of hydromel, for it beareth an egg to the breadth of a sixpence."

Metheglin of the vintage 1717 would probably prove anything but palatable to the simpler tastes of to-day, for the worthy knight's receipt includes thyme, eglantine, sweet marjoram, rosemary, ginger, cinnamon, cloves and pepper, in fairly liberal proportions.

" Miodomel," a beverage prepared by the monks of St Basil, at Sokal in Poland, is a species of mead flavoured with hops. Dobrogost Chylinski in his " Beekeeper's Manual " (1845) thus describes the method of its preparation :—

" To twenty-four gallons of water put twelve gallons of honey and twelve pounds of hops ;[1] boil them together on a *very slow fire*, till the whole is reduced one third. Care must be taken that the fire be not too strong, yet the heat must increase gradually ; from a sudden and excessive heat, a burnt taste will be communicated to it. From the boiler empty it into a large tub or barrel, which must be deposited in a warm place during eight days, so as to undergo the process of fermentation ; afterwards it must be filtered through a woollen filter into a barrel, and placed in a cellar for use. The older it is, the better and stronger it becomes. After it has been twelve months in the cellar it may be bottled, and kept for years."

Chylinski extols the virtues of miodomel as improving and restoring the power of digestion, effective against gout and rheumatism, and a most excellent remedy for measles.

Krupnik, another Polish drink, consists (says Chylinski) of neat whisky boiled with honey, and should be drunk warm, especially in winter.

[1] Early British mead usually contained hops.—H. R.

CHAPTER V

COTTAGERS AND BEEKEEPING

BEEKEEPING on modern lines is a rural industry especially adapted to cottagers. The profits resulting from the produce of a few hives if properly managed should pay the rent and leave a trifle over. As an example, a journeyman carpenter says that after paying all expenses, "I generally manage in the worst of seasons to buy the family a suit of clothes all round,"—the family consisting of eight all told.

A Yorkshire farmer's wife made in 1900 £10 profit from her bees, "and that with a lot less worry and work than with any of their other stock." And, enlarging upon this latter point, she quaintly remarks, "there is no 'sitting up' with them, and they always swarm in the middle of a fine day."

There is also a moral side to the question, and the cottager who embarks upon beekeeping soon comes under its spell, his intelligence quickens and his interests are enlarged. Cotton emphasises this point somewhat in his preface to that most charming of all bee-books, "My Bee Book." "Again," he says, speaking of the labourer, "his bee hives are close to his cottage door; he will learn to like their sweet music better than the dry squeaking of a pot-house fiddle, and he may listen to it in the free open air with his wife and children about him. They will be to him a countless family. He will be sure to love them if he cares for them, and they will love him too and repay all his pains."

The chief difficulty confronting the ordinary cottager

desirous of replacing his straw skeps with the up-to-date appliances of modern beekeeping is one of £ s. d., the expenditure of thirty or even fifteen shillings in one lump sum in the purchase of appliances being well nigh out of the question.

To overcome this difficulty it is suggested that the transition from ancient to modern bee culture might be  undertaken gradually. As a first step, one or more supers for skeps might be purchased. These cost 3s. 6d. and contain fifteen 1 lb. sections each. The bottom of the crate is covered with a board in the centre of which is a round hole fitted with queen-excluder zinc.

SECTIONAL SUPER FOR SKEP

If there is no hole in the top of the skep one should be cut by means of a sharp knife of the same size as the hole in the crate. A little clay or mortar should be placed around this hole and the crate pressed down firmly, taking care that it be perfectly level. On each side of the crate a chain with a wire pin is fixed. These pins being pushed into the sides of the skep prevent the super being blown off. The whole is covered with a span roof, this again being secured by means of pins back and front.

Here then the cottager by a moderate expenditure enhances considerably the value of his honey, and the increased profits so gained should help him to advance another stage by purchasing a bar-frame hive. A really good hive may be had for the modest sum of eight shillings and sixpence. Having purchased one hive, he might, if at all handy, use it as a pattern for making others of his own. As an instance of what can be done

in this way, a correspondent in the *British Bee Journal*
gives the following details regarding the timber used
by him for the construction of four hives.

"I. Three bacon boxes which held long 'singed
sides' of bacon—the boards being over 40 ins. long—
and two lengths each. Cost of each 10d.   2. Three
full-sized Ceylon Tea chests, used for roof and internal
fittings. Cost of each 6d.   3. Four Swiss Milk cases;
these made the plinths, etc. Cost of each 3d."

The boxes were all carefully taken to pieces, nails
straightened and used again, the cost being thus 1s. 3d.
per hive; paint and putty 1s. 4d. Total 2s. 7d. These
hives were made to take ten "Standard" frames, two
supers, with stand (no legs presumably) floor board,
extra super and top to match.

When extracted honey is sought for, the cost of an
extractor is a serious item, but this necessity might be
a joint-stock possession.

In round numbers the value of the honey imported
into this country every year amounts to £27,000. The
amount of honey represented by this figure could easily
be produced in this country, and that by the cottager
class to their manifest advantage.

CHAPTER VI

The Spring.—Paradoxical as it may sound, the best treatment for bees in the early spring is to let them severely alone. Should there be any doubt as to a sufficiency of stores, a cake of candy (previously warmed) should be carefully placed over the cluster as expeditiously as possible, and without unduly disturbing the bees.

Towards the latter end of March, should the season be favourable, éxamine each colony carefully, making sure of the presence of a queen. Contract the brood-nest by means of the division-boards, confining the bees in as close a space as possible, so as to utilise to the full the natural heat of the cluster. A quilt of American cloth, glazed side downwards, supplemented with plenty of felt or carpet quilts, will still further assist in the conservation of the warmth.

Provide water and artificial pollen if necessary. Syrup-feeding may be commenced.

Especially at this period, take every precaution against robbing, by keeping the entrances contracted. Untimely manipulation, and the spilling of syrup about the apiary, are often instrumental in setting up robbing.

Stocks found queenless should be united to one possessing a queen.

In May, the winter coverings may be removed, the brood-chamber lifted bodily, and the floor-board replaced with a clean one. Or the old one may be swept by an assistant (carefully collecting and burning all the sweepings), and the brood-chamber replaced.

All hives not in use should be carefully scrubbed out in every part with a weak solution of carbolic acid and stored ready for swarms, should such unfortunately issue.

Have all supers in readiness for the first honey flow. Time is precious when honey really begins to come in in earnest.

The Summer.—The details of summer management will be mostly found fully described in the body of this book under the heading of surplus honey. When, towards the end of the summer, honey is becoming scarce, keep a sharp look-out to prevent robbing. Start slow feeding, with the object of keeping up breeding as long as possible. The autumn-raised bees will prove most useful in carrying the stock through the winter.

In the middle of September, commence to feed up all stocks with less than 20 lbs. of sealed stores by means of the rapid feeder, and before finally packing down for the winter, replace the quilt next the frames with a clean one, having previously cleaned the top bars as much as possible from adhering propolis.

Don't forget to provide winter passage-ways across the tops of the frames.

Should the hives be at all likely to experience the full force of the winter gales, large stones or bricks should be placed on the roofs, to prevent them being blown over.

The combs in the brood-nest should be renewed about every third year. Use a full sheet of foundation and a new frame and ends. The old comb can go in the wax smelter, but the old frame should be burnt at once. Take care to replace only one or two of the combs at a time : those in the centre of the brood-nest will be found to age the soonest, whilst the outside combs may not require replacing for many years. The methodical beekeeper will mark the date of each comb on the projecting lug of the top bar.

Apply to the leading bee-appliance manufacturers for copies of their catalogues. They will be found to be profusely illustrated, and many of them contain capital hints on bee management, together with a calendar of operations. An occasional order (either direct or through their agents) for foundation, sections, etc., will suggest itself to the conscientious recipient.

Take care to order all goods required for the forth-coming season as early as possible, so as to ensure prompt delivery. Some dealers offer a special discount on all goods ordered before a certain date, usually about the middle of March.

A card should be nailed in the roof of each hive, or a slate may be conveniently kept (in the case of the W.B.C. hive) in the space between the hive proper and the outer case, upon which can be jotted down notes relative to the condition of each colony. The systematic posting of these notes in a note-book much simplifies bee-work, especially if a large number of stocks are kept.

CHAPTER VII

CONCLUSION

THERE are three points essential to successful bee-keeping that the novice should always bear in mind.

Firstly. Union is strength. It is no use whatever trying to build up colonies hopelessly weak. All such should be promptly united.

Secondly. All feeding is profitable. Do not grudge the cost and trouble of feeding stocks in the spring and autumn. Attention to this point is essential.

Thirdly. To prevent swarming, give room somewhat in advance of the bee requirements, and, *above all*, do not tolerate queens more than two years old. It is a question if it would not be better to re-queen at the close of each season.

In order to profit by the experience of others, and also with the object of keeping up to date, the reader is recommended to peruse the *British Bee Journal*, a weekly publication devoted entirely to apiculture, or the monthly *Bee-keepers' Record*. Through the medium of either journal queries relating to bee-keeping are gratuitously answered.

Almost every county in England has its Bee-keeping Association, each association employing an expert, who visits, at least once a year each member, giving advice and practical help when needed. The subscription to such association is merely nominal, and in addition to the benefits personally derived, there is the knowledge that these associations are doing much towards fostering a neglected rural industry.

There is also a British Beekeepers' Association, with which the county associations are affiliated, the latest enterprise of which has been the successful launching of a system of insurance whereby, for a small annual premium, beekeepers may insure against accidents caused by their bees outside their apiaries. Particulars of this may be obtained through the Secretaries of the various County Associations, or from the Secretary of the British Beekeepers' Association, 12 Hanover Square, London, W.

INDEX

A.

Acid, carbolic, 71, 105.
Ants, 112.
Apiary, arrangement of the, 14.
Appliances, cleaning, 106.
   ,,   list of, 21.
Arrangement of apiary, 14.
Artificial feeding, 60.
   ,,   pollen, 62, 122.
   ,,   swarming, 58.
Association, British Beekeepers',
  126.

B.

Bacillus Alvei, 107.
Bar-frame hive, 22.
Bee appliances, list of, 21.
Bee, Brown, 12, 13.
   ,,   Carniolan, 13.
   ,,   Cyprian, 13.
   ,,   drone, 5.
   ,,   English, 12.
   ,,   escape, the " Porter," 78.
   ,,   life of, 6.
   ,,   Ligurian, 12.
   ,,   queen, description of, 4.
   ,,   Syrian, 13.
   ,,   worker, 5.
Beekeepers' Association, 126.
   ,,   insurance for, 126.
Beekeepers' Record, 125.
Beekeeping, cottagers and, 119.
Bees, cross-breeding of, 13.
   ,,   diseases of, 107.
   ,,   distances of flight, 20.
   ,,   domestic economy of, 7.

Bees, enemies of, 112.
   ,,   and flowers, 1.
   ,,   flowers fertilised by, 2.
   ,,   gorged with honey, 7.
   ,,   hybrid, 13.
   ,,   natural history of, 4.
   ,,   and pollen, 3.
   ,,   removing, 114.
   ,,   how to rid supers of, 78.
   ,,   varieties of, 12.
   ,,   water for, 15.
   ,,   wax, 7.
Bell heather, 18.
Berlepsch, Baron von, 22.
Beta-napthol, 61.
Birds, 113.
Blind louse, 113.
Block for frames, 36.
Board, floor, 24.
Boards, crown, 42.
   ,,   division, 36.
Bottles, Cowan travelling crate
  for, 103.
Bottles, honey, 104.
Brood, chamber, 24, 32.
   ,,   chilled, 108.
   ,,   drone, how distinguished, 9.
   ,,   foul, 107.
   ,,   lift, 24.
Brice, H. W., method of queen
  rearing, 89.
British Bee Journal, 125.
Brown bee, 12, 13.
Butler, 70.

C.

Canadian feeder, 63.
Candy, 64.

Made in the USA
Columbia, SC
08 October 2020

22355764R00105